SKY AND PURPOSE IN PREHISTORIC MALTA

Sky and Purpose in Prehistoric Malta:
Sun, Moon, and Stars at the Temples of Mnajdra

SOPHIA CENTRE MASTER MONOGRAPHS: VOLUME 2

Sophia Centre for the Study of Cosmology in Culture
University of Wales Trinity St David
Jennifer Zahrt, General Editor

SKY AND PURPOSE
IN PREHISTORIC MALTA

Sun, Moon, and Stars at the
Temples of Mnajdra

by Tore Lomsdalen

SOPHIA CENTRE PRESS

© TORE LOMSDALEN 2014

First published by Sophia Centre Press in 2014.

All rights reserved. No part of this publication may be
reproduced or utilised in any form or by any means, electronic
or mechanical, including photocopying, recording, or by any
information storage and retrieval system, without permission
in writing from the Publishers.

SOPHIA CENTRE PRESS
University of Wales Trinity Saint David
Ceredigion, Wales SA48 7ED, United Kingdom
www.sophiacentrepress.com

PUBLISHER'S CATALOGING-IN-PUBLICATION
(PROVIDED BY CASSIDY CATALOGUING SERVICES, INC.)

Names:	Lomsdalen, Tore, author.
Title:	Sky and purpose in prehistoric Malta : sun, moon, and stars at the temples of Mnajdra / by Tore Lomsdalen.
Description:	Ceredigion, Wales, United Kingdom : Sophia Centre Press, 2014. | Series: Sophia Centre master monographs ; volume 2 | Includes bibliographical references and index.
Identifiers:	ISBN: 978-1907767418 (paperback) | 978-1-907767-64-7 (ebook)
Subjects:	LCSH: Archaeoastronomy—Malta. | Astronomy, Ancient—Malta. | Malta—Antiquities.
Classification:	LCC: GN799.A8 L66 2014 | DDC: 520.94585—dc23

ISBN 978-1-907767-41-8
British Library Cataloguing in Publication Data.
A catalogue card for this book is available from the
British Library.

Book Design by Joseph Uccello.

Printed in the UK by LightningSource.

ABSTRACT

The exploration of the Mediterranean seascape goes back to the foragers of the early Holocene period around the ninth millennium BCE. However there is no secure evidence of human settlement of the Maltese Archipelago before the end of the sixth millennium BCE. Approximately one thousand years later, the unique style of megalithic constructions that later became known as the Temple Period commenced.

The temple period in Malta goes from the Early Neolithic (4,100 BCE) until the Early Bronze Age (2,500 BCE) and then suddenly came into an unexplainable decline. However, when it comes to the Mnajdra complex the core time frame is Ġgantija (3,600-3,000 BCE) and the Tarxien (3,000-2,500 BCE) Phases. The temple complex consists of three distinct structures: the small trefoil temple, the north (or middle) temple and the south (or lower) temple. Each temple has a clearly defined orientation. The central axis of the south temple is aligned towards the sunrise at the spring and autumn equinoxes. The middle temple is oriented towards south-east and the winter solstitial sunrise, whereas the trefoil temple has a south-west orientation towards the islet of Filfla.

Based on site observations, horizon astronomy, architecture and archaeological findings, this book proposes a redefined constructional chronology of the Mnajdra Temple and that part of it may intentionally have been constructed by the temple builders to observe the rising of the sun at equinoctial and solstitial time periods.

*To my parents,
my children Patrick and Miriam,
and my grandchild Ian.*

ACKNOWLEDGEMENTS

A sincere thanks and gratitude to all the following who have been involved in, supportive and contributing to this research: George Barbaro Sant, Dr. Bernadette Brady, Marcia Butchart, Konrad Camilleri, Dr. Nick Campion, Daniel Cilia, Clive Cortis, John Cox, Nic Galea, Anna Grima, Dr. Reuben Grima, Prof. Kim Malville, Saviour Sacco, Dr. Fabio Silva, Katya Stroud, Mario Vassallo, Prof. Frank Ventura and Dr. Jenn Zahrt. Furthermore I am deeply grateful to the Ministry of Tourism, Culture and Environment and Heritage Malta for giving me access to the temples.

TABLE OF CONTENTS

Acknowledgments v
List of Figures xii
Foreword xvii

Chapter 1
Introduction 1
   1.1   AIM 1
   1.2   MALTA'S 'BEST KEPT SECRET' 3
   1.3   COSMOLOGY AND ASTRONOMY IN MALTA 6

Chapter 2
Maltese Prehistory: A Literature Review 13
   2.1   THE NEOLITHIC DIFFUSION IN A EUROPEAN CONTEXT 13
   2.2   THE COLONISATION OF MALTA AND THE
          EARLY NEOLITHIC 17
   2.3   THE TEMPLE PERIOD 24
   2.4   THE MNAJDRA TEMPLE COMPLEX 31
          2.4.1   HISTORY OF SITE RESEARCH 33
          2.4.2   LANDSCAPE 36
          2.4.3   DESCRIPTION OF THE TEMPLES 39
          2.4.4   BUILDING SEQUENCE 62

Chapter 3
Maltese Cosmology and Astronomy: A Literature Review 65

3.1 COSMOLOGY AND LANDSCAPE 65
3.2 TEMPLE PERIOD ASTRONOMY...70
3.3 MNAJDRA AND THE COSMOS 82

Chapter 4
Methodology 91
4.1 SITE VISITS 91
4.2 SURVEYING..92
4.3 ASTRONOMICAL OBSERVATION AND PHOTOGRAPHY 94
4.4 EXPERIMENTAL ARCHAEOLOGY.......................................97
4.5 PHENOMENOLOGY 102

Chapter 5
Results... 105
5.1 ARCHAEOASTRONOMICAL SURVEY OF
     MNAJDRA EAST AND MIDDLE TEMPLES.......................... 107
     5.1.1   MNAJDRA EAST 107
     5.1.2   MNAJDRA MIDDLE 108
5.2 ARCHAEOASTRONOMICAL SURVEY OF
     MNAJDRA SOUTH TEMPLE... 112
     5.2.1   MAIN ENTRANCE... 113
     5.2.2   HORIZON POSTHOLES 116
     5.2.3   CROSS-QUARTER AND EIGHTH DAYS 122
     5.2.4   ORACLE HOLES 128
     5.2.5   OTHER POSSIBLE ALIGNMENTS 129

Chapter 6
Discussion.. 135
6.1 MALTESE ARCHAEOASTRONOMY................................... 135
6.2 INTENTIONALITY BEHIND MNAJDRA............................. 137

	6.2.1	OFFSET ILLUMINATION AND THE LIGHT/DARK DICHOTOMY 138
	6.2.2	AN ASTRONOMICAL INTENTION BEHIND THE ORACLE HOLES IN MNAJDRA SOUTH 142
	6.2.3	THE SOLSTITIAL POSTHOLES 143
	6.2.4	THE CONSTRUCTION SEQUENCE OF MNAJDRA .. 145

Chapter 7
Conclusion 160

Appendix I 163
Appendix II 186
Appendix III 208
Appendix IV: Glossary 221
Bibliography 224
Index 233

LIST OF FIGURES AND TABLES

Figures:

2.1 .....Neolithic diffusion in the Mediterranean basin 17
2.2 .....Major temple sites of the Maltese archipelago 25
2.3 .....Arial photo of the Mnajdra complex 32
2.4 .....Plan of the Mnajdra complex 36
2.5 .....Façade of Mnajdra complex 38–39
2.6 .....East Temple seen from its southern entrance 40
2.7 .....Broken portal entrance to the Middle Temple 43
2.8 .....North apse in Room 7 43
2.9 .....Engraved temple façade and the Magrr
          temple site slab 44
2.10 ...Back altar of the Middle Temple 47
2.11 ...South-west niche of the Middle Temple 47
2.12 ...Rope hole in front of Mnajdra South entrance 49
2.13 ...Well-preserved entrance and central
          corridor of Mnajdra South 49
2.14 ...Room 1 in Mnajdra South 51
2.15 ...Entrance to Room 3 from Room, anno
          2014 and 1868 53

2.16...Altars in Room 3...55
2.17...Entrance to Room 2 and its back altar 57
2.18...Photo of Mnajdra South before restoration work..........59
2.19...First known photos ever taken of Mnajdra, 1868......60–1
3.1 .....Spirals and animal representations at
         the Taxien Temples..67
3.2 .....The orientations of the temple axes................................75
3.3 .....A schematic showing temple orientations 76–7
3.4 .....The Trefoil temple seen from the back altar
         towards Filfla Island..79
3.5 .....Two tally stones in the East Temple.................................81
3.6 .....WSSR through the broken portal
         entrance of the Middle Temple 83
3.7 .....Plan of WSSR and MJLS alignments at
         the Middle Temple 84
3.8 .....Slit image of illumination of Mnajdra South at
         solstice and equinox 86
3.9 .....Plan of equinoctial and solstitial illumination
         of Mnajdra South 88
4.1 .....Photographing the EQSR and SSSR at
         Mnajdra South..95
4.2 .....Jupiter rising at dec. 0° on 26 June 2010
         at 23:19 UT..99
4.3 .....Portable poles used for alignment at
         WSSR, Mnajdra Middle... 101
5.1 .....East Temple with its orientations 108
5.2 .....Middle Temple with alignments 109
5.3 .....WSSR seen from the corner of the left
         altar in Room 7... 111
5.4 .....Manjdra South with its eastern orientation 113

5.5 ..... Horizon from the main entrance of
   Mnajdra South...114–5
5.6 ..... EQSR observations in Mnajdra South.......................116–7
5.7 ..... WSSR observations in Mnajdra South 118–9
5.8 ..... SSSR observations in Mnajdra South........................120–1
5.9 ..... The main entrance of Mnajdra South.......................... 123
5.10... Rising of the moon on 26 June 2010 124
5.11... SSSR indicating the actual sunrise in
   relation to the posthole.. 125
5.12... Constructed image of slit illumination
   of sunrise throughout the year...................................... 126
5.13... Alignments from oracles holes
   towards WSSR and SSSR 128
5.14... WSSR illumination from oracle hole in Room 5............. 130
5.15... Alignments from back altar Room 2 131
5.16... Alignments from the left and right
   altars in Room 1 132
5.17... Alignments from Room 3 133
5.18... Mnajdra South and the sunrise
   alignments I investigated 133
6.1 ..... EQSR illumination of back altar, Mnajdra South 139
6.2 ..... WSSR illumination in the Middle Temple 140
6.3 ..... An artistic impression of a possible
   roofed Mnajdra South 141
6.4 ..... The suggested first three building stages
   of Mnajdra South 146–7
6.5 ..... The suggested fourth and fifth building
   stages of Mnajdra South 148–9
6.6 ..... Room 5 of Mnajdra South with its dressed wall 150

Tables:

2.1 ..... *Maltese ancient chronological history* 24
5.1 ..... *Comparative field research of azimuths, altitudes, and declinations of the Maltese Prehistoric Temples* 106
5.2 ..... *Cross-Quarter and Eighth Days in 2012* 127

FOREWORD: *Sky and Purpose in Prehistoric Malta*

The prehistoric temples of the Maltese Islands were mentioned in the literature for the first time in 1647. Impressed by their massive construction with huge stones, the author attributed them to an undated mythical age when giants inhabited these islands. In spite of this early exposure, interest in the temples increased only very gradually. In the eighteenth and early nineteenth centuries several scholars visited the sites, described the half-buried structures, produced illustrations of them and gave opinions about their builders. By the end of the eighteenth century, the structures were attributed to the Phoenicians, the legendary builders who used the Maltese harbours in their journeys across the Mediterranean Sea and settled here in about 700 BCE and possibly earlier.

Eventually in the early 1820s, the interiors of the contiguous twin temples at Ġgantija in Gozo—then known as the Giant's Tower—were cleared of soil, dirt and other debris. The clearing and excavation of the other major sites followed in the next hundred years. However, although the architecture and building technique of the temples became clearer and more impressive, many scholars still considered the Maltese structures as rude monuments. They also thought that the spiral decorations on some of the stones were simply rough copies of similar decorations in Mycenean monuments. This wrong impression was corrected much later in the 1960s when, in a re-writing of prehistory, radiocarbon dating showed that the earliest Maltese temples were older than 5,500 years and that the temple culture lasted for about a thousand years. Indeed it was then realised that the temples were much older than any other free-standing stone monuments anywhere and their source could not be attributed to any other culture. This realisation led UNESCO to recognise the megalithic temples of Malta as World Heritage sites and to acknowledge that each of the main temples is 'a unique architectural masterpiece which would be immensely impressive at any date, given the limited resources of the builders, but is quite staggering when taken with the extraordinarily early dates attributed to them'.

While it took so long to recognise the international significance of the temples, it is surprising that the first speculations about the possible connection between the temples and the sky began as early as 1840 when J. G. Vance was commissioned to clear the interior of the complex temple at Ħaġar Qim. In his report Vance commented that several observations induced him to believe that the site was devoted to the worship of the heavenly bodies—the sun, the moon and the stars. He also conjectured that Ħaġar Qim and the nearby temple site at Mnajdra were dedicated to the sun and the moon respectively although he did not mention any specific supporting evidence. This lead was apparently lost for a long time and it was only when Italian archaeologist Luigi Ugolini visited the temples in 1934 that it was taken up afresh. Ugolini noted that the temples have a special orientation and there seems to be some connection between the central axes of the temples and the motion of the stars or other objects we see in the sky. The eminent archaeologist John Evans disagreed, and in 1959 he commented that 'Neither the individual temples, nor the temples as a group seem to have a consistent orientation, although most of the entrances face towards SE or SW...therefore it seems that the orientation was not important. There is no sign of any special interest in any object that we see in the sky'. In spite of this negative opinion of one of the foremost experts on the Maltese temples, the idea that there could be a connection between astronomy and the temple orientations flourished in the 1970s. This probably happened because of the great publicity given at that time to the work by Gerald Hawkins on Stonehenge and the meticulous surveys by Alexander Thom on other megalithic structures in England and elsewhere. The first tentative work on temple orientations in Malta was that by photographer Gerald Formosa, which was published in 1975, who noticed how shafts of sunlight penetrated the temple at Ħaġar Qim at the summer solstice. A year later, a local newspaper [in Maltese] published two articles by Paul Micallef on the orientations of the temples of Ħaġar Qim and Mnajdra and their presumed connection with the sun.

The first comprehensive survey of the orientations of the temples then took place in the winter of 1979–1980 with the measurement of twenty axes corresponding to the central corridors of the temples and another six axes of side entrances. The data showed that the temples had a clear signature, as the builders had preferred to set the entrances to face a direction between southeast and southwest. The declinations were then checked against the declinations of the rising and setting of the sun and the moon at the turning points and the equinoxes as well as the declinations of stars brighter than the second magnitude. Several possible alignments with the sun, the moon at its major standstill, and the stars were discussed. The most remarkable was that of the Mnajdra South temple which happened to be aligned with sunrise at the equinox. This alignment immediately raised questions on whether the temple builders actually intended to align the temple with the equinox or whether it was a chance alignment. This happened because not only is it hard to believe that the builders arrived at the abstract concept of equinox at such an early date, but it is also difficult to accept that the builders had the necessary background knowledge to find the position of sunrise at the equinox with good accuracy on site. The subsequent investigations which attempted to answer the question of intentionality are discussed in detail in the present publication, so it is superfluous to deal with them here. However, it is necessary to mention another comprehensive survey of temple orientations which took place in 1991 and which basically confirmed the findings of 1980. In addition, this new survey noted that the Mnajdra South temple could have been aligned with the Pleiades, which had the same declination as the equinox around 3000 BCE when the temple was built. In support of this alignment, a series of parallel rows of drilled holes in the East Mnajdra temple were interpreted as a record of dates of heliacal risings of stars or star groups starting with the Pleiades. This development suggested that the intention of the builders could have been to direct their temple towards the heliacal rise of the Pleiades rather than towards equinox sunrise.

In this book Tore Lomsdalen ably reviews the relevant archaeological context and the various archaeoastronomy surveys of the temples of Malta that have been carried out since 1980. He then focuses on the main aim of this study, which is to produce an answer to the question of the intentionality of the alignment of the Mnajdra South temple. In doing so he draws on his extensive fieldwork in Malta and on the analysis of other works to introduce new evidence and insights to support his conclusions. Although the strength of the results of archaeoastronomical studies cannot compare with the strength of the results of radiocarbon ^{14}C dating, hopefully, this work will contribute to a wider recognition of importance of the sky in the cosmology of the Neolithic temple builders of the Maltese Islands and to an appreciation of its international significance.

FRANK VENTURA
University of Malta
15 February 2014

CHAPTER 1

INTRODUCTION

1.1 AIM

The aim of this book is to investigate whether the Neolithic Mnajdra Temples on the island of Malta were deliberately built as sacred sites for religious worship and to pay respect to the power of the cosmos. It will particularly examine solar alignments at the equinoxes and solstices to establish if they were built and oriented intentionally as devices for calibrating time and the seasons. Based on the temples' architecture, archaeological observations and astronomical alignments, I will propose a redefined constructional chronology for the Mnajdra compound. The questions of whether astronomically significant features of the temples were intentional or epiphenomenal will be addressed.

Further, it will be argued that Malta's Mnajdra South Temple might be the oldest known site that qualifies as a

Neolithic device intentionally constructed to cover the entire path of the rising sun throughout the year. It predates two other apparently similar sites: Taosi in China by one millennium and Chankillo in Peru by about two millennia. In addition, my research shows that the Mnajdra South Temple has clearly demarcated areas, more so than the two above-mentioned sites, which are illuminated at sunrise; at the time of the equinoxes the central corridor is completely illuminated, whereas at the solstices, quarter- and eighth-days there is a cross-jamb illumination of two vertical orthostats.[1] This argument stands in contrast to Ruggles':

> Around the world we have many sites that are aligned on equinox or solstice sunrise, like Stonehenge. But there are only two sites known in the world that I'm aware of, where you have a device that seems to cover a whole arc. They've both been discovered in the last five years or so. One is Taosi, in China, and the other Chankillo, in Peru.[2]

Ruggles' suggestion that Chankillo was built as a solar observation device has been contested by J. McKim Malville on the grounds that the observation point, or 'Observation Building' as named by Ivan Ghezzi and Ruggles, is not clearly established and that a horizon calendar appears to have

---

1 John Cox and Tore Lomsdalen, 'Prehistoric Cosmology: Observations of Moonrise and Sunrise from Ancient Temples in Malta and Gozo', *Journal of Cosmology* 9 (2010).
2 Clive Ruggles, 'Heavenly Power in Worldly Hands: Ancient Sky Perceptions and Social Control', (paper presented at the European Society for Astronomy in Culture [SEAC], Gilching, Germany, 2010).

been an unintended consequence of the initial design of the thirteen towers.³ However, David Pankenier *et al.* suggest that Taosi is a 'rudimentary horizontal calendar and accurate determination of the length of a solar year'.⁴

1.2 MALTA'S 'BEST KEPT SECRET'

The Maltese archipelago is situated roughly in the middle of the Mediterranean basin and consists of two main islands, Malta and Gozo, covering a total of 316 square kilometres. It lies about 80 km south of Sicily, 255 km north of Libya and 815 km west of Crete.⁵ The earliest evidence of human presence on the islands goes back to the early Neolithic (about 5,000 BCE).⁶ Around a thousand years later, 'these religious people started with the erection of megalithic temples which at that time were as unique as

---

3 J. McKim Malville, 'Astronomy and Ceremony at Chankillo: An Andean Perspective', in *Archaeoastronomy and Ethnoastronomy: Building Bridges between Cultures*, ed. Clive L. N. Ruggles (Cambridge: Cambridge University Press, 2011), pp. 15–61; Iván Ghezzi and Clive Ruggles, 'The Social and Ritual Context of Horizon Astronomical Observations at Chankillo', *Proceedings of the International Astronomical Union* 7 (2011): pp. 144–53.
4 David W. Pankenier *et al.*, 'The Xiangfen, Taosi Site: A Chinese Neolithic "Observatory"?', *Archaeologia Baltica, Astronomy and Cosmology in Folk Traditions and Cultural Heritage* 10 (Klaipeda: University of Klaipeda, 2008).
5 David H. Trump, *Malta: Prehistory and Temples* (Malta: Midsea Books, 2002), p. 15.
6 Trump, *Malta: Prehistory*, p. 23.

they are today'.[7] Originally there may have been approximately forty temples on the islands of which about twenty remain today in various conditions.[8] Some excavations were undertaken in the late nineteenth century, but most archaeological work was conducted in the twentieth century, with some buildings being reconstructed or rebuilt.[9] The Maltese temples do not appear as isolated monuments but are frequently found in groups, often paired or even clumped together.[10] It is during the Temple Period that the Mnajdra complex was built and developed. This complex consists of three individual buildings with archaeological evidence spanning two chronological phases: the Ġgantija Phase (3,600 to 3,000 BCE) and the Tarxien Phase (3,000 to 2,500 BCE).[11]

Although the prehistoric structures are characterised as temples, the question of the definition of a temple arises. Douglas Davies suggests that temples are sacred buildings, echoing something of the central sacred space of a

---

7 Themistocles Zammit and Ing. Karl Mayrhofer, *The Prehistoric Temples of Malta and Gozo: A Description by Prof. Sir Themistocles Zammit* (Malta: Ing. Karl Mayrhofer, 1995), p. 5.

8 Rowland Parker and Michael Rubinsetin, *Malta's Ancient Temples and Ruts* (London: The Institute for Cultural Research, 1988), p. 2.

9 John Cox, 'Observations of Far-Southerly Moonrise from Hagar Qim, Ta' Hagrat and Ggantija Temples from May 2005 to June 2007', *Cosmology Across Cultures*, ASP Conference Series 409 (San Francisco: Astronomical Society of the Pacific, 2009): p. 344.

10 A. Bonanno *et al.*, 'Monuments in an Island Society: The Maltese Context', *World Archaeology* 22, no. 2 (1990): p. 193.

11 Anthony Pace, 'The Building of Megalithic Malta', in *Malta before History*, ed. Daniel Cilia (Malta: Miranda Publishers, 2004), p. 18.

religion.[12] According to Mircea Eliade, a temple is the 'Center of the World', where the sacred manifests in space and opens communication between the cosmic planes (earth and heaven), bringing the world into existence through the sanctification of space, equivalent to cosmogony.[13] According to Giulio Magli, to name the Maltese megalithic structures as 'temples' is circumstantial, due to a lack of reliable evidence as to why these monuments were erected.[14] On the other hand, Caroline Malone *et al.* suggests that Malta provides one of the best documented cases of prehistoric rituals, as there is even evidence of a building with religious functions, a 'shrine', that predates the temples.[15] Robin Skeates agrees that the temples 'probably did function, at least in part, as sacred places for worship'.[16] Today, most scholars seem to acknowledge the structures as temples or ritual monuments; David Trump even states that, 'there can now be very little argument but that they

---

12 Douglas Davies, 'Introduction: Raising the Issues', in *Sacred Place, Themes in Religious Studies* (London: Continuum, 1994), p. 5.
13 Mircea Eliade, *The Sacred and the Profane: The Nature of Religion* (Orlando: Harcourt, Inc., 1959), pp. 43, 63.
14 Giulio Magli, *Mysteries and Discoveries of Archaeoastronomy from Giza to Easter Island* (New York: Copernicus Books, 2009), p. 49.
15 Caroline Malone *et al.*, 'Introduction', in *Cult in Context: Reconsidering Ritual in Archaeology*, eds. David A. Barrowclough and Caroline Malone (Oxford: Oxbow Books, 2007), p. 3; David Trump, 'Maltese Temple Cult: The Antecedents', in *Cult in Context: Reconsidering Ritual in Archaeology*, eds. David A. Barrowclough and Caroline Malone (Oxford: Oxbow Books, 2007), p. 14.
16 Robin Skeates, *An Archaeology of the Senses* (Oxford: Oxford University Press, 2010), p. 156.

really were temples'.[17] As temple construction reached its highest flourishing, it went into a sudden and unexplainable decline around 2,500 BCE. According to Magli, both the explosive start and the mysterious end of the complex giant megalith construction period raise many questions. After the Temple Period, Malta seems to have been colonised by a new Bronze Age civilization which possessed none of the skills in masonry or architectural ability of their predecessors.[18]

1.3 COSMOLOGY AND ASTRONOMY IN MALTA

The location of the temples, often built on slopes and facing south, appears to have been important to their builders, according to Reuben Grima.[19] A relationship to the sea seems to prevail with a marked preference for locations with maritime connectivity, suggesting that the temples might have been a ceremonial gateway between land, sea, and the outside world. This could very well have been the framework of the islanders' cosmology.

Today's cosmologists ask the same questions that people

---

17 David H. Trump, *Malta: An Archaeological Guide* (London: Faber and Faber Ltd., 1972), p. 24; Christopher Tilley, *The Materiality of Stone: Explorations in Landscape Phenomenology: 1* (Oxford: Berg, 2004), p. 92.

18 Magli, *Mysteries*, pp. 48–49.

19 Reuben Grima, 'Landscape and Ritual in Late Neolithic Malta', in *Cult in Context: Reconsidering Ritual in Archaeology*, eds. David A. Barrowclough and Caroline Malone (Oxford: Oxbow Books, 2007), pp. 36–40.

have asked for thousands of years.[20] Among those are questions involving the sky. Nicholas Campion links cosmology with astronomy when he suggests that 'the sky is an essential part of human existence. Landscapes do not exist without skyscapes'.[21] Human behaviour may be guided by the belief that life on earth is an imitation of celestial events, and temples are often considered a microcosm of the universe that incarnate and express cosmological beliefs.[22] The recognition of the cyclicality of celestial movements and the deliberate marking of them makes ancient astronomies observable in the ethnographic and archaeological record.

In this context the discipline of archaeoastronomy mingles with the study of prehistory. Ruggles defines archaeoastronomy as: 'The study of human perceptions and actions relating to the sky', whereas Malville emphasises that the challenge is to understand the ancient sky watchers and to be able to see the heavens through their eyes.[23]

The sun moves at an average rate of 1° (about twice its diameter as seen from the Earth) per day along the horizon at its point of rising; as Malville states, 'the oscillation of the sun between summer and winter sunrise points along the

---

20 J. McKim Malville, *A Guide to Prehistoric Astronomy in the Southwest* (Boulder, CO.: Johnson Books, 2008), p. 3.

21 Nicholas Campion, 'Locating Archaeoastronomy within Academia', in *TAG Liverpool 2012, 34th Annual Conference of the Theoretical Archaeology Group* (University of Liverpool, 2012).

22 David H. Kelly and Eugene F. Milone, *Exploring Ancient Skies: An Encyclopedic Survey of Arcaeoastronomy* (New York: Springer, 2005), p. 2.

23 Ruggles, 'Heavenly'; Malville, *Prehistoric*, p. 3.

horizon gave ancient astronomers a convenient calendar'.[24] However, the sun does not move at a uniform rate along the horizon; it slows down as it nears each solstice, like a pendulum reaching the end of its swing.[25] At the solstices, the sun's rising point on the horizon for a northern hemisphere observer reaches its most northerly position in the summer and its most southerly position in the winter. At the equinoxes the sun rises due east. According to David H. Kelly and Eugene F. Milone, in Paleolithic and Neolithic cultures the azimuth (the compass bearing measured clockwise in degrees from true north) of the rise or set of a celestial object along the horizon was probably marked by sightlines directed to these points.[26] The purpose and layout of these sightlines, in combination with a back-sight or observing position, and their focus on astronomical events on the horizon constitute what is known as 'horizon astronomy', an important component of archaeoastronomy.

Luigi M. Ugolini was, in 1934, the first person to relate the temples to astronomy and orientation.[27] He mentioned the fragment of the Tal-Qadi stone as an astrological stone.[28] Themistocles Zammit, in 1929, suggested that a pattern of five holes in the forecourt of the Tarxien Temple represented an image of the constellation of Crux (South-

---

24 Malville, *Prehistoric*, p. 36.
25 Malville, *Prehistoric*, p. 37.
26 Kelly and Milone, *Exploring Ancient Skies*, p. 166.
27 Luigi M. Ugolini, *Malta: Origini Della Civilta Mediterranea* (Malta: La Libreria dello Stato, 1934), p. 128. Translations from Italian my own.
28 Ugolini, *Origini*, p. 181.

ern Cross).²⁹ George Agius and Frank Ventura confirmed that this constellation could, in fact, be clearly seen from Malta during the temple period.³⁰

Nevertheless, it was not until later that the temples aroused the archaeoastronomical interest of both scholars and general enthusiasts. The first scientific survey of Maltese temple orientations and their astronomical significance was performed by Agius and Ventura in the 1970s/80s.³¹ In 1990 it was brought to the attention of the public in the more popular booklet, *Mnajdra, Prehistoric Temple: A Calendar in Stone*, published by Paul I. Micallef.³² Analyses of twenty-six temple axis azimuths indicate that the builders had a southeast and southwest preference in their temple orientations.³³ John Cox proposes that the temple orientations of Gozo and Malta show a consistency, which suggests that some temples might intentionally

---

29 Themistocles Zammit, *The Neolithic Temples of Hal-Tarxien, Malta: A Short Description of the Monuments with Plan and Illustrations*, 3rd ed. (Valletta, Malta: Empire Press, 1929), p. 13.

30 George Agius and Frank Ventura, 'Investigation into the Possible Astronomical Alignments of the Copper Age Temples in Malta', *Archaeoastronomy* 4, no. 1(1981): p. 16.

31 George Agius and Frank Ventura, *Investigation into the Possible Astronomical Alignments of the Copper Age Temples in Malta* (Malta: University Press, 1980).

32 Paul I. Micallef, *Mnajdra Prehistoric Temple: A Calendar in Stone* (Malta: Union Print, 1990).

33 Agius and Ventura, *Investigation*, pp. 8–9; Giorgia Foderà Serio *et al.*, 'The Orientations of the Temples of Malta', *Journal for the History of Astronomy* 23, no. 2 (1992): pp. 109–16.

have been constructed to face particular directions.³⁴ On this note, Mario Vassallo, an amateur archaeoastronomer who for a long time has investigated the Maltese temples, concludes that the temple people seemed to consider the winter solstice sunrise to be the more important time for sunrise observations; several temples have cross-jamb demarcated illumination through the temple entrances onto a left-hand altar inside.³⁵ This argument is sustained by Klaus Albrecht, who suggests that the winter solstice appears to have been selected as the most important time of year to symbolise immortalisation of their religious ideas in architectural form.³⁶

The South Temple at Mnajdra is unique; its main axis is oriented towards the east, where the sun rises at the equinoxes. According to Ventura this could have been pure chance, referring to the statistical 'parable' of the random throwing of twenty-six matches.³⁷ If the matches' direction is analogous to the orientation of the twenty-six temples, it is not surprising that one of the matches lands in an easterly direction.³⁸

In archaeoastronomy it is important to distinguish be-

---

34 John Cox, 'The Orientations of Prehistoric Temples in Malta and Gozo', *Archaeoastronomy* 16 (2001): p. 36.

35 Mario Vassallo, 'Sun Worship and the Magnificent Megalithic Temples of the Maltese Islands', *The Sunday Times of Malta* (23 January 2000), pp.40–41; (30 January 2000), pp. 44–45; (6 February 2000), pp. 36–37.

36 Klaus Albrecht, *Malta's Temples: Alignment and Religious Motives* (Postdam: Sven Näther Verlag, 2007), p. 11.

37 Tore Lomsdalen, 'A Talk with Frank Ventura: Astronomical Observations Related to Maltese Prehistoric Temples', printed in Appendix I in this volume.

38 Lomsdalen, 'A Talk with Frank Ventura', in Appendix I, of this volume.

tween 'orientation' and 'alignment'. All straight lines have an orientation. Thus, all structures from which one can define a straight line—for example the main axis of a temple or any given straight wall—are oriented. Some might be oriented 'towards' something; a stone row might be oriented towards a mountain on the horizon but this might be a coincidence, not an intentional aim of the row's builders. If the feature was intentional, one can then speak of an 'alignment' with the mountain. This topographic example can be easily extended to alignments with celestial events. In simple terms, an alignment implies intentionality, whereas an orientation does not. According to Ruggles this distinction is important when seeking evidence of astronomical concerns reflected in prehistoric monumental architecture.[39] Thus, this research seeks to discern whether the orientation of the Mnajdra South Temple is, or is not, an alignment.

This chapter has introduced the aim, intention, and possible outcome of my research. It further presents what is, according to Ventura *et al.*, 'perhaps the best-kept secret in Mediterranean archaeology', referring to the prehistoric Maltese temples.[40] The temple culture has been considered within a sacred, cosmological, and astronomical context. The next chapter will explore details of Maltese prehistory.

---

39 Clive Ruggles, *Ancient Astronomy: An Encyclopedia of Cosmologies and Myth* (Oxford: Abc-Clio, 2005), p. 8.
40 Frank Ventura, 'Temple Orientations', in *Malta before History*, ed. Daniel Cilia (Malta: Miranda Publisher, 2004), p. 107.

CHAPTER 2

MALTESE PREHISTORY: A LITERATURE REVIEW

THIS CHAPTER PROVIDES a literature review of Maltese prehistory. It examines the European transition model from hunter-gatherer to agricultural and settled societies in relation to Maltese settlement and the development of an indigenous temple culture, and it describes prior research done on the Mnajdra Temple Complex.

2.1 THE NEOLITHIC DIFFUSION IN A EUROPEAN CONTEXT

Around 50,000 years ago, the first groups of behaviourally modern humans began to move into the Mediterranean basin, starting in the Levant and northeast Africa.[1] At the end of the last Ice Age and the Younger Dryas, about 12,000 BCE, the Palaeolithic peoples of Europe were living as hunter-gatherers in small mobile groups.[2] These types of societies continued throughout the Mesolithic period for about another 6,000 years. The transition from hunting and gathering to farming is one of the most significant developments in prehistory.[3]

By the second half of the eighth millennium BCE definite indications of agro-pastoralism appear on the Mediterranean island of Cyprus, suggesting a material culture very similar to Pre-Pottery Neolithic (PPNB) villagers in Southwest Asia.[4] In the fifth millennium the same transition is visible in the archaeological record of Western Europe.[5] This transition in subsistence strategy was interpreted as a revolution by Gordon Childe, who hypothesised that agriculture was introduced to Europe by Neolithic peoples from the Near East in search of new land, spreading in waves across the continent.[6] According to Jacques

---

1 Cyprian Broodbank, 'The Origin and Early Development of Mediterranean Maritime Activity', *Journal of Mediterranean Archaeology* 19, no. 2 (2006): p. 205.
2 Steven Mithen, *After the Ice: A Global Human History 20,000–5000 BC* (London: Phoenix, 2003), pp. 3–4.
3 Robert Leighton, *Sicily before History: An Archaeological Survey from the Palaeolithic to the Iron Age* (London: Duckworth, 1999), p. 34.
4 Graeme Barker, *The Agricultural Revolution in Prehistory: Why Did Foragers Become Farmers?* (Oxford: Oxford University Press, 2006), p. 346.
5 Barker, *Prehistory*, p. 325.
6 Vere Gordon Childe, *Man Makes Himself* (Bradford-on-Avon: Moonraker Press, 1981), p. 68.

Cauvin, this thesis has lost more and more ground.[7] Based on the quality of evidence, Barker adds that it is 'particularly unlikely' that Neolithic immigrants came in waves from Southwest Asia into Europe.[8] However, the transition from Mesolithic to Neolithic culture is a vast, complex and repetitive scenario based more on multidirectional mobility than on a series of major unidirectional migrations.[9]

Currently, there are two possible transition models: one of colonisation by farmers and one of acculturation by Mesolithic foragers who adopted Neolithic customs, eventually becoming Neolithic themselves.[10] Fabio Silva concludes, 'at least in Atlantic Europe, native Neolithisation, or acculturation, seems to have been the most important process'.[11] Nevertheless, in the central and eastern Mediterranean a potential Levantine Neolithic influence cannot be excluded as seafarers were already travelling that part of the world from the eleventh millennium BCE.[12]

Besides agriculture, the Neolithic period distinguished itself further from the Mesolithic as people began to construct stone monuments.[13] These vary from the more simple menhirs to passage graves, stone circles and other mas-

---

7 Jacques Cauvin, *The Birth of the Gods and the Origins of Agriculture* (Cambridge: Cambridge University Press, 2000), p. 137.
8 Barker, *Prehistory*, p. 390.
9 Barker, *Prehistory*, p. 378.
10 Fabio Silva, 'Cosmologies in Transition: Continuity, Innovation and Transformation in Neolithic Europe' (MA thesis, University of Wales, Trinity Saint David, 2012), p. 40.
11 Silva, 'Transition', p. 32.
12 Broodbank, 'Maritime', p. 212.
13 Silva, 'Transition', p. 3.

sive megalithic structures. Some of these were, and still are, oriented towards cyclical celestial events.[14] According to Christopher Tilley, the perception of permanence and anchoring of monuments served to make the connection between people and the land for the first time.[15] This visibly brought the presence of the ancestral past to consciousness. Eliade maintains that stone is a manifestation of power, and its sacred value is due to it being part of something beyond itself, suggesting that its never-changing existence, grandeur, and hardness manifest a predictability which transcends human uncertainty.

Eliade further argues that the change from a nomadic to a settled lifestyle implied vital decisions involving the entire community, coupled with acts that presupposed an existential choice: one is prepared to live in a universe by creating it. To settle into a territory is a process of consecration.[16] Julian Thomas, however, remarks that the foragers' seasonal movements between resources also implied movement between landmarks which are understood in a cosmological context: rocks, trees, rivers, and mountains could be linked to special events, spirits, deities, or abstract qualities.[17] Often, there is a powerful ancestral link both to the monuments themselves and the cyclical movements of people.

---

14 Silva, 'Transition', p. 5.
15 Christopher Tilley, *A Phenomenology of Landscape* (Oxford: Berg Publishers, 1994), p. 202.
16 Eliade, *The Sacred and the Profane*, p. 34.
17 Julian Thomas, *Understanding the Neolithic* (Oxon, UK: Routledge, 1999), p. 35.

2.2 THE COLONISATION OF MALTA AND THE EARLY NEOLITHIC

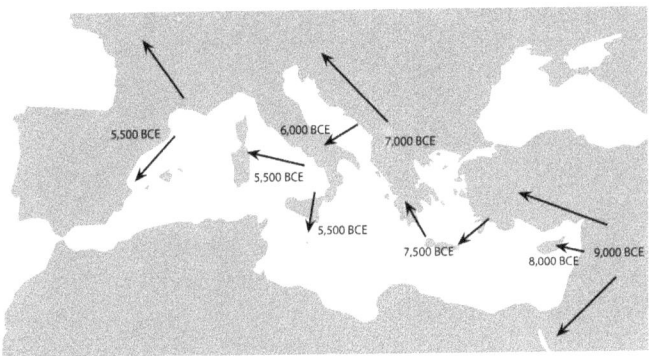

FIGURE 2.1. Neolithic diffusion in the Mediterranean basin, reaching Malta. Adapted from Trump.[18]

During the Last Glacial Maximum, LGM, (around 21,000–18,000 BCE) several parts of the Mediterranean and the Balkans acted as refuge areas for humans moving from more extreme conditions.[19] As the climate became milder, hunter-gatherer populations expanded rapidly through the Mediterranean Basin; according to Barker 'it is very likely that these early Holocene foragers made use of sea-going craft capable of coping with Mediterranean currents,

---

18 Trump, *Malta: Prehistory*, p. 54.
19 Cyprian Broodbank, *The Making of Middle Sea: A History of the Mediterranean from the Beginning to the Emergence of the Classical World* (London: Thames & Hudson, 2013), p. 116.

tides, and winds'.[20] Eighteen thousand years ago Malta and the Aegadian islands were joined to Sicily. It was only by 9,000 BCE that the Mediterranean coastlines had more or less reached their present configuration.[21]

Malta and Gozo are among the most remote islands in the Mediterranean and not directly visible from any mainland, although from Malta, on clear days, Sicily's Mount Etna can just about be seen on the horizon.[22] However, under optimal atmospheric conditions the crossing between Malta and Sicily may be undertaken without ever losing sight of land.[23] Mark Patton considers Malta an atypical island for early colonisation due to its smallness and distance from any mainland. Empirical studies suggest that the larger an island is, and the closer it is to a mainland, the earlier it is likely to be colonised.[24] Three large Mediterranean islands—Corsica, Sardinia and Cyprus—possess evidence of colonisation from the Early Holocene period. The size of these islands could support permanent hunter-gatherer populations, something that Malta could not offer.[25]

For a long time it was believed that the Phoenicians

---

20 Graeme Barker, 'Agriculture, Pastoralism, and Mediterranean Landscapes in Prehistory', in *The Archaeology of Mediterranean Prehistory*, eds. Emma Blake and A. Bernard Knapp (Oxford: Wiley-Blackwell, 2005), p. 49.

21 Mark Patton, *Islands in Time: Island Sociogeography and Mediterranean Prehistory* (London: Routledge, 1996), p. 36.

22 Patton, *Islands*, p. 104.

23 Reuben Grima, 'The Prehistoric Islandscape', in *The Maritime History of Malta: The First Millennia*, eds. Charkes Cini and Jonathan Borg (Malta: Salesians of Don Bosco and Heritage Malta, 2011), p. 15.

24 Patton, *Islands*, p. 37.

25 Patton, *Islands*, p. 59–61.

were the first settlers on Malta.[26] However, in 1929, Zammit rejected these arguments and dated the megalithic period to 'at least 3000 BC'.[27] The first scientific evidence of colonisation of the Maltese Archipelago was based on uncalibrated radiocarbon dating and set at, or slightly before, 4,000 BCE.[28] Later re-calibration using the now-standard tree-ring calibration curve, moved these dates back one thousand years, indicating that colonisation goes back to 5,266–4,846 BCE.[29] Geological evidence suggests that the Maltese land was more fertile and hospitable to the first settlers than it is today.[30] According to Trump, the earliest immigrants were farmers; but, to reach Malta, they also had to be sailors with boats large and seaworthy enough to transport domestic animals such as sheep, goats, pigs, and cattle. They also brought with them cereals like barley, primitive forms of wheat, emmer, and lentils; remains of all these species types of grain have been found in the bottom layers of the Skorba Temple site.[31] At this site, pottery from the Ghar Dalam Phase (5,200–4,500 BCE) with decorations identical to those of Stentinello-type—commonly found in Sicily—has been found.[32] Gozo also used and pro-

---

26 A. A. Caruana, *Report on the Phoenician and Roman Antiquities in the Group of the Islands of Malta* (Malta: Government Printing Office, 1882).

27 Themistocles Zammit, *Malta: The Islands and Their History*, 2nd ed. (Valletta, Malta: The Malta Herald, 1929), pp. 21–22.

28 Trump, *Archaeological*, p. 20.

29 Trump, *Malta: Prehistory*, p. 23.

30 Caroline Malone *et al.*, 'The Death Cults of Prehistoric Malta', *Scientific American* 269, no. 6 (December, 1993): pp. 110–17, here p. 116.

31 Trump, *Archaeological*, p. 20.

32 Trump, *Archaeological*, p. 20.

duced pottery similar to that found in Ghar Dalam.[33]

Regarding the origins of the Maltese settlers, Bernard A. Vassallo suggests either Sicily—even though its rich archaeological heritage possesses nothing like the Maltese temples—or North Africa.[34] J. D. Evans finds the connection to the African megaliths vague and maintains that they all seem to be younger than the Maltese temples.[35] However, Malville hypothesises that the temple builders could have descended from, or been influenced by, the megalith builders of Nabta Playa in southern Egypt.[36] Due to drought and dramatic climate changes the Nabta Playa nomads moved further north around 4,000 BCE, which matches the start date of the megalithic culture in Malta. Another colonization theory offered by Anton and Simon Mifsud maintains that the first settlers came from the fertile Franco-Cantabrian regions, and reached the Maltese islands during the last millennia of the Ice Age.[37] According to Simon Stoddard *et al.*, current archaeological evidence suggests that early Maltese immigrants arrived from the north and colonised a previously unoccupied archipelago.[38] John Robb suggests that the passage from Sicily to Malta under Neo-

---

33 Patton, *Islands*, p. 51.
34 Bernard A. Vassallo, *Prehistoric Malta, Europe and North Africa* (Valletta, Malta: Allied Publications Ltd, 2007), p. xiv.
35 J. D. Evans, *Malta: Ancient People and Places*, ed. Daniel Glyn, Ancient Peoples and Places (London: Thames and Hudson, 1959), p. 25.
36 J. McKim Malville, Email, 26 May, 2010.
37 Anton Mifsud and Simon Mifsud, *Dossier Malta: Evidence for the Magdalenian* (Malta: Propprint, 1997), p. 170.
38 Simon Stoddart et al., 'Cult in an Island Society: Prehisrtoric Malta in the Tarxien Period', *Cambridge Archaeological Journal* 3, no. 1 (1993): p. 6.

lithic navigational conditions, with small boats or canoes being rowed or sailed, would be feasible in one to three days.[39] That possibility was proved valid by Patrick Brydone's trip in 1780; with two companions, three servants and several hired boatmen, he sailed from Sicily to Malta in a small, oar-propelled boat.[40] At a little after 9:00 PM the boat embarked from Sicily, at about 2:00 AM discovered the island of Malta and, in less than three hours more, reached the city of 'Valetta'. This experimental sea voyage demonstrated that one can row and sail a small oar-propelled craft from Sicily to Malta in less than 24 hours. Cox suggests the safest months for passage (at present) seem to be May and June, as a current runs north to south past the Maltese islands; however, this pattern of currents would make a return passage much more difficult, even hazardous.[41] In fact, Stoddard suggests that the colonisation of Malta and Gozo could have been seasonal in its intention.[42] Grima suggests that it is more than likely that the decision to set up permanent settlements on the Maltese islands was preceded by several exploratory journeys.[43] According to Trump, attempts to establish an island settlement prior to 5,000 BCE are pure guesswork, although he maintains that people were sailing and trading in the Mediterranean by

---

39 John Robb, 'Island Identities: Ritual, Travel and the Creation of Difference in Neolithic Malta', *European Journal of Archaeeology* 4, no. 2 (2001): p. 187.
40 Patrick Brydone, *Tour through Sicily and Mata: In a Series of Letters to William Beckford* (London: Forgotten Books 2012, 1806), pp. 177–79.
41 Cox, 'Orientations', p. 36.
42 Stoddart *et al.*, 'Cult', p. 6.
43 Grima, 'Maritime', p. 13.

8,000 BCE, well before farming was introduced to the area; therefore, earlier colonisation of Malta 'was by no means impossible'.[44]

The early immigrants would have maintained close contact with Sicily and beyond, as certain raw materials, such as obsidian and flint, do not occur naturally in Malta and yet are found in their Early Neolithic records.[45] The settlers' pottery quickly evolved its own unique style, different from anything produced elsewhere. At about 4,000 BCE, there is a marked change, which ends the Ghar Dalam Phase and begins the Zebbug Phase (4,100-3,800 BCE).[46] In this phase there appears to have been continued cultural contact with Sicily, mainland Italy and even the Alps; obsidian from Lipari and Pantelleria, flint from Sicily and greenstone axes from the crystalline rocks of the Sila area in Calabria were found.[47] Some of the stone axes could also have come from Alpine sources.[48] As for Maltese exports, no recognisable objects have yet been identified elsewhere.[49] However, Sebastiano Tusa claims that in the Licata territory in Southern Sicily, within a river a short distance from the sea, a number of Maltese pottery shards of Tarxien style were found.[50] Robb argues that Malta may have been a sort of

---

44 Trump, *Malta: Prehistory*, pp. 23–24.
45 Robb, 'Identities', p. 187.
46 Trump, *Archaeological*, p. 20.
47 Stoddart *et al.*, 'Cult', p. 7.
48 Patton, *Islands*, p. 151.
49 J. D. Evans, 'Island Archaeology in the Mediterranean: Problems and Opportunities', *World Archaeology* 9, no. 1 (1977): p. 20.
50 Sebastiano Tusa, *La Sicilia Nella Preistoria* (Palermo: Sellerio, 1999), pp. 406–7. Translated from Italian by Lomsdalen.

trade cul-de-sac, a terminal point in a chain of circulation and re-working of art and ceremonial objects that resulted in a continuing importation of primary goods.[51] Maltese prehistoric society's cultural uniqueness may possibly be related to the more hazardous sea voyage from Malta to Sicily.[52]

Radiocarbon dating suggests that about two thousand years after colonisation, Malta experienced a population pressure; however, no archaeological evidence suggests unrest or violence of any kind and no recognizable weapons have ever been found.[53] Possibly the cultural isolation stimulated an internal energy channelled into the development of its own monuments; a period of stressed isolation may have created a crisis in social-cultural functioning as well, resulting in a collapse of their initial cultural heritage between the fourth and third millennium.[54] This environmental change may have been the origin of the next prehistoric phase in Maltese society, namely The Temple Period.

---

51 Robb, 'Identities', p. 188.
52 Cox, 'Orientations', p. 36.
53 Evans, 'Archaeology', p. 24.
54 Stoddart *et al.*, 'Cult', p. 5.

2.3 THE TEMPLE PERIOD

NEOLITHIC	
GHAR DALAM	5,000–4,300 BCE
GREY SKORBA	4,500–4,400 BCE
RED SKORBA	4,400–4,100 BCE
TEMPLE PERIOD	
ZEBUGG	4,100–3,700 BCE
MGARR	3,800–3,600 BCE
GGANTIJA	3,600–3,200 BCE
SAFLINI	3,300–3,000 BCE
TARXIEN	3,150–2,500 BCE
BRONZE AGE	
TARXIEN CEMETRY	2,400–1,500 BCE
BORG IN-NADUR	1,500–700 BCE
BAHRIJA	900–700 BCE
PHOENICIAN	
PHOENICIAN	70–550 BCE
PUNIC	550–218 BCE
ROMAN	218 BCE–330 CE

TABLE 2.1. Malta's ancient chronological history.[55]

What is generally known as the Maltese Temple Period ranges from 4,100 to 2,500 BCE.[56] However, indications of temple and religious rituals may go back some centuries to the Red Skorba Phase that immediately precedes the Tem-

---

55 Trump, *Malta: Prehistory*, p. 55.
56 David Trump, 'Dating Malta's Prehistory', in *Malta before History*, ed. Daniel Cilia (Malta: Miranda Pubishers, 2004), p. 230.

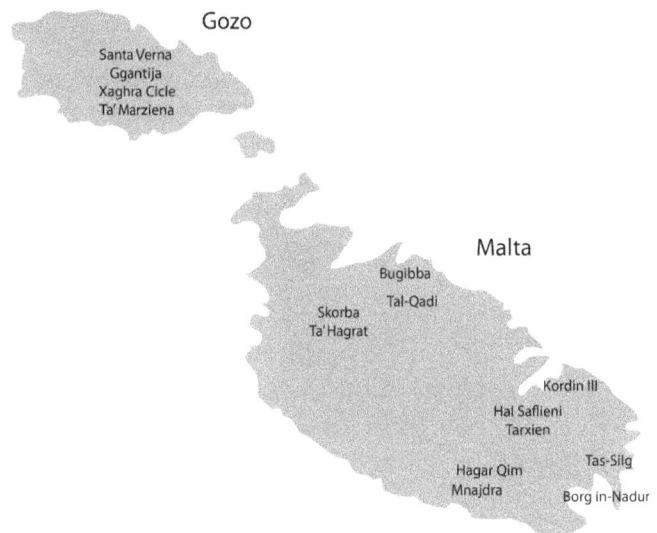

FIGURE 2.2. Major temple sites of the Maltese archipelago.[57]

ple Period.[58] In Skorba, a number of terracotta and stone figurines, animal bones expressing phallic symbols, treated goat skulls, bone-worked tarsals of cows, and an almost equal number of animal bones and pottery were found in a hut-like building which Trump describes as a 'Shrine'.[59] This shrine and its contents are the earliest extant examples of a probable distinction between sacred and every-

---

57 Magli, *Mysteries*, p. 48.
58 David H. Trump, *Skorba* (Oxford: University Press, 1966), p. 10.
59 Trump, *Skorba*, pp. 11–14.

day life.[60] Figurines from this period, some with vaginal exposure, are the earliest known examples of the graphic female representations that have become iconic symbols of Maltese prehistory.[61] Together with phallic iconography, these may be the first indications of fertility cults and ritual in Malta.[62]

The Temple Period proper begins with the Zebbug Phase (4,100–3,800 BCE), although this still precedes the building of the megalithic temples. The first anthropomorphic figurines, retrieved at the Xaghra Stone Circle, an underground necropolis in Gozo, date from this phase.[63] Perhaps population expansion, enlargement of family groups or symbolic reasons left no space for individualism in this communal underworld. People were buried together in family groups encompassing several generations, in the company of common ancestors, in rock-cut tombs and underground burial chambers where simple, collective burial and death rites were conducted.[64] Red ochre was predominantly used for decoration; flint objects and green stone axes were common grave goods. These early burial practices and rites carry many similarities to those found on Sicily.[65] In 1955 another site dated to this phase was excavated. At the Xemxija Heights, five well preserved, rock-cut tombs were

---

60 Pace, 'Building', p. 24.
61 Stuart Piggott, *Ancient Europe: From the Beginnings of Agriculture to Classical Antiquity* (Chicago: Aldine Publishing Company, 1965), pp. 114–15.
62 Malone *et al.*, 'Cults', p. 18.
63 Stoddart *et al.*, 'Cult', p. 7.
64 Evans, 'Archaeology', pp. 23–24.
65 Malone *et al.*, 'Cults', p. 22.

found with an odd, arbitrary, curvaceous and kidney-shaped plan. Evans maintains that these constructions 'proved to be the key to the whole development of both the rock-cut and the built monuments of the Maltese islands'.[66] This argument is maintained by Stuart Piggott who sees a development from rock-cut tombs to the building of huge stone monuments as part of a particular island cult.[67]

The relatively short intermediate Mġarr Phase (3,800–3,600 BCE) refers to the type of pottery found at the Ta' Ħaġrat Temple site near the village of Mġarr, about one kilometre westward from the Skorba site. The building at this site looks more primitive and predates the more elaborate constructions from the subsequent Ġgantija Phase (3,600–3,000 BCE).[68] According to Evans, Ta' Ħaġrat (which means 'the stone heap') is the 'most vital site for the understanding of the development of the temple-culture in Malta'.[69] This is also the period connected to ancient cults and religions in which fertility worship may have been an important component.[70] According to Caroline Malone *et al.*, Malta and Gozo were still relatively fertile and not overpopulated; half a millennium later, the archipelago was shaken by major environmental change and degradation as soil erosion occurred; at the same time, an increase in population caused problems, which led to Malta's next

---

66 Evans, *Malta*, pp. 88–89.
67 Piggott, *Ancient*, p. 97.
68 John D. Evans, *The Prehistoric Antiquities of the Maltese Islands: A Survey* (London: The Athlone Press University of London, 1971), pp. 33–34.
69 Evans, *Malta*, p. 26.
70 Malone *et al.*, 'Cults', pp. 21–22.

prehistoric phase.[71]

The Ġgantija (3,600-3,000 BCE) and the succeeding Tarxien Phase (3,000-2,500 BCE) are the two major construction phases of the Temple Period on the Maltese islands. The Ġgantija Phase finds a sudden wave of megalithic construction and an abundance of ceramics with few parallels outside Malta. This period sees the apparently sudden appearance of a complex form of architecture with highly developed plans, special designs, advanced engineering and the building of several megalithic monuments, usually in clusters, in various locations throughout Malta and Gozo. Anthony Pace points out that this was a single cultural process with no monumental similarity elsewhere in Europe.[72] Its cultural distinctiveness was not restricted to pottery and ceramic, but extended to the ritualisation of the landscape on which the temples and megalithic monuments were founded.[73] Religious influence, social hierarchy and control over the population might have been influencing factors, leading into the Tarxien Phase.[74]

The Tarxien Phase (3,000-2,500 BCE) is the pivotal and final phase of the Temple Period.[75] Malta became increasingly dominated by what Colin Renfrew calls a 'chiefdom society' where cult and ritual specialists (a priesthood) enjoyed enormous prestige, power and control, both eco-

---

71 Malone et al., 'Cults', p. 22.
72 Pace, 'Building', pp. 28–29.
73 Stoddart et al., 'Cult', p. 7.
74 Malone et al., 'Cults', p. 22.
75 Malone et al., 'Cults', p. 22.

nomic and social.⁷⁶ A vast amount of time and effort was put into the construction of temples and artistic activities where the dead were honoured through ceremonial feasts and rituals. Malone *et al.* argues that this was done at the cost of neglecting or putting little effort into building housing quarters and villages, agricultural development or environmental improvement.⁷⁷ During this period many of the existing temples seem to have been rebuilt, extended or enlarged and the complex Tarxien temple, the nearby necropolis and the 'temple in the negative'—the Hal Saflieni Hypogeum—were built.⁷⁸ The extension of Mnajdra to the middle temple was also built in this period and completed the Mnajdra compound as it stands today.⁷⁹

During this period as many as thirty temples, possibly more, might have existed on Malta and Gozo.⁸⁰ Renfrew estimates that, during the Temple Period, Malta had a total population of eleven thousand, about two thousand people for each of the six geographic temple regions that he identified.⁸¹ Grima maintains that this estimate is too high, as it is based more on the agricultural landscape of the medieval period, involving techniques that do not appear until the Bronze Age. Instead, he estimates a maximum of around

---

76 Colin Renfrew, *Before Civilization: The Radiocarbon Revolution and Prehistoric Europe* (London: Pimlico, 1973), pp. 170–74.
77 Malone *et al.*, 'Cults', p. 22.
78 Anthony Pace, 'The Sites', in *Malta before History*, ed. Daniel Cilia (Malta: Miranda Publishers, 2004).
79 Evans, *Antiquities*, p. 103.
80 Robb, 'Identities', p. 178.
81 Renfrew, *Civilization*, p. 169.

five to six thousand people living in the Maltese archipelago (see Appendix III). Daniel Clark, who calculated the required labour for the construction of the temples, concludes that it was well within the population's capabilities and resources, and that no great strain was put on the inhabitants. He suggests that how the population was organised was probably of more significance in the achievement of such impressive results than logistics alone.[82] On this note, Evans suggests that Maltese society was a strongly structured hierarchy with a dominant elite whose prestige gave them power to organise the necessary labour force to build the temples. The drive to make temples would be fostered by intergroup rivalry, a stimulus to erect ever larger and more impressive structures temples, thus creating a peaceful outlet for tensions in an isolated island society under considerable pressure.[83]

In general the population seems to have been healthy, with few dental problems or other detectable illnesses; the anatomical and genetic evidence indicates that there was little or no change in the genetic makeup of the early Maltese inhabitants.[84] This again supports the idea that changes in their customs were not the result of foreign immigration. Robb suggests that during the Temple Period, Malta was regarded as isolated, inward-turning and basically rejecting of contact with the outside world.[85] This could

---

82 Daniel Clark, 'Building Logistics', in *Malta Before Hitory*, ed. Daniel Cilia (Malta: Miranda Publishers, 2004), p. 377.

83 Evans, 'Archaeology', p. 23.

84 Malone *et al.*, 'Cults', p. 21.

85 Robb, 'Identities', p. 186.

again be due to the difficult ocean currents in the direction of Sicily.[86] Little imported pottery has been found and the locally produced pots are remarkably different from those of contemporary Italian and Sicilian styles, apart from the examples previously mentioned.

By the end of the Tarxien Phase temple building had reached a culmination. Five hundred years later (by 2,000 BCE) the archaeological record features a very different religious practice, favouring cremation burials. The cult represented by the figurines and rituals related to the living and the dead was completely abandoned.[87] According to Magli, the society came to a halt for no apparent reason, 'encouraging the impression that the civilization that built the temples just disappeared'.[88] Soon afterwards Malta seems to have been colonised by a population with all the typical features of the Bronze Age, with nothing resembling the masonry and architectural capacities and abilities of their predecessors.[89]

2.4 THE MNAJDRA TEMPLE COMPLEX

Today the Mnajdra complex consists of three distinct temple structures accessed through a common, paved, open, central forecourt, about thirty metres across (Fig. 2.4).

---

86 Cox, 'Orientations', p. 36.
87 Malone *et al.*, 'Cults', pp. 22–23.
88 Magli, *Mysteries*, pp. 48–49.
89 Magli, *Mysteries*, pp. 48–49.

Through the modern entrance to the site, the first temple encountered is the small eastern trefoil temple, oriented towards the southwest. The next building is the north or Mnajdra Middle Temple with its spectacular portal entrance, now partly broken. The last structure is the lower or Mnajdra South Temple with its centrally placed entrance and concave façade.

FIGURE 2.3. Aerial photo of the Mnajdra complex, with courtesy of Daniel Cilia.

2.4.1 HISTORY OF SITE RESEARCH

The name Mnajdra suggests a Semitic origin, a derivation from *'mandra'*, meaning a small enclosure or (a more popular usage of the word) *'a mess'*.[90] Old photographs taken prior to the restoration work do suggest a general sense of disorder (see Figs. 2.18 and 2.19). Giovanni Francesco Abela is credited with being the first person to explicitly mention the prehistoric remains of the temples on Malta.[91] Abela, who thought the temples were built by giants, did not mention Mnajdra specifically. However he did mention its next door neighbour, Ħaġar Qim, and the small island 'Folfola' (today known as Filfla), just off the coast south of Mnajdra, as a religious place.[92] That the Filfla islet served as a marker or focal point—symbolic and physical—for both Ħaġar Qim and Mnajdra East is further addressed by England who compared the altar-shaped islet to the profile of a bull's horn, an image that may have attracted the attention of Neolithic peoples.[93]

According to Evans, Mnajdra was first excavated in 1840 by Lenormant; however, no report was ever published.[94] The first reported excavation seems to have been made by Fergusson who, in 1872, gave a fuller description of the site

---

90 Pace, 'Sites', p. 127.

91 Evans, *Antiquities*, p. 3.

92 Franscesco Abela, *Malta Illustrata: Della Descrittione Di Malta* (Malta: Paolo Bonacota, 1647; repr. Facsimile Edition, Midsea Books Ltd, 1984), p. 145.

93 Richard England, 'A Space-Time Genealogy', in *Malta before History*, ed. Daniel Cilia (Malta: Miranda Publishers, 2004), p. 413.

94 Evans, *Antiquities*, p. 95.

together with a rough sketched plan.[95] In 1882 Caruana also published a report on Mnajdra; however, according to Evans, 'his illustrations were simply poor reproductions of Fergusson's', and until then, Mnajdra was generally treated merely as an appendix to Ħaġar Qim.[96] It was not until 1901 that a more academic and systematic study of the temples and artefact finds was made by Albert Mayr, a German archaeologist.[97] Mnajdra was again excavated in 1913 by Ashby, the Director of the British School in Rome, who was the first to carry out substantial restoration work on the site. Ashby was the first to describe the small trefoil building and to re-erect the fallen central West pillar.[98] Ashby was also the first to present an accurate plan of the Mnajdra complex.[99] Although Fergusson does indicate the back niche of the small trefoil temple in his plan of the Mnajdra complex, it remains an open question as to why that building is neither indicated nor mentioned in Mayr's excavation report.[100] Following up on the work of Mayr and Ashby, Zammit published his own investigations with plans and illustrations in 1927, and claimed that Mnajdra consisted of two structures, with the small trefoil Temple as a mere

---

95 J. Fergusson, *Rude Stone Monuments in All Countries: Their Age and Uses* (London: John Murray, 1872), pp. 418–22.
96 Caruana, *Phoenician* pp. 14–17; Evans, *Antiquities*, p. 95.
97 Albert Mayr, *Die vorgeschichtlichen Denkmäler von Malta* (München: Verlag der k. Akademie, 1901), pp. 656–64.
98 Thomas Ashby et al., *Excavations in 1908–11 in Various Megalithic Buildings in Malta and Gozo* (London: Macmillan & Co., 1913), p. 91.
99 Ashby et al., *Excavations*, Plate XX.
100 Fergusson, *Monuments*, p. 419; Mayr, *Denkmäler*, pp. 656–64.

subsidiary building.[101] In the 1930s studies of the Mnajdra temple were concerned with the structure of the buildings and brought up the question of roofing.[102] Both Ugolini, and Ceschi suggest that the temples were roofed; the latter even made a study and illustration of the theoretical roofing of the South Temple.[103] This possibility is supported by Trump when he refers to 'the now-missing roof'.[104] Evans also advocates that the temples may have been roofed.[105] The question of roofing may be strengthened by retrieval of a small slab formed as a roofed temple at the Ta' Ħaġrat Temple.[106] It was not until after the Second World War in 1948 that the National Museum began the work of clearing, tidying and restoring Mnajdra and it once more became of interest for archaeological studies in the 1950s. During the next decades extensive research was conducted by John D. Evans, as well as by the previous Curator of Archaeology in the Malta National Museum, David H. Trump.[107]

---

101 Themistocles Zammit, *The Copper Age Temples of Ħaġar Qim and Mnajdra: With Plans and Illustrations* (Valletta, Malta: Facsimile Edition, 1927), pp. 24–25.
102 Katya Stroud, *Ħaġar Qim & Mnajdra Prehistoric Temples* (Malta: Heritage Books, 2010), p. 15.
103 Ugolini, *Origini*, p. 57.
104 Trump, *Malta: Prehistory*, p. 150.
105 Evans, *Malta*, pp. 126–27.
106 Alex Torpiano, 'The Construction of the Megalithic Temples', in *Malta Before History*, ed. Daniel Cilia (Malta: Miranda Publishers, 2004), p. 348.
107 Evans, *Antiquities*, pp. 95–104; Trump, *Archaeological*, p. 97.

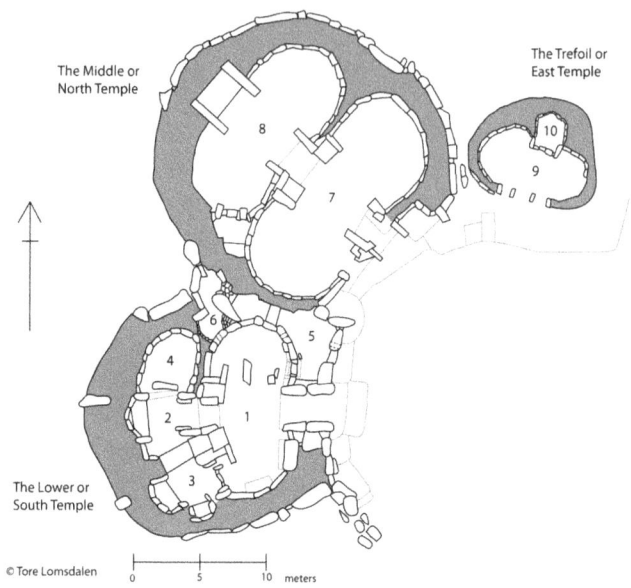

FIGURE 2.4. The three temples of the Mnajdra complex with numbering of the rooms and apses. Plan by Lomsdalen, adapted from Evans.[108]

2.4.2 LANDSCAPE

Mnajdra is situated on the southeastern part of Malta on a slope leading to the sea. One's first impression is that the landscape where Mnajdra was built is barren and inhospitable; however, it offered all the resources necessary for

---

108 Evans, *Antiquities*, Plan 20A.

a community 5,500 years ago.[109] According to Grima, it is a clear example of how temple builders took advantage of the availability of nearby material. Although mainly built from the harder Lower Coralline Limestone found on cliffs dropping into the sea, the softer Globigerina Limestone is also available, 200 metres or less, from the site.[110] The Lower Coralline Limestone, being more resistant to weathering, was used in the outer walls, whereas the softer Globigerina Limestone was used to create special finishes and decorated interiors. Because of this the Mnajdra complex contains some of the better-preserved temples and may never have completely disappeared under accumulated debris since the time it went out of use.

According to Evans the façade of Mnajdra South Temple gives an archaic impression and the middle temple was never altered after its completion.[111] Various construction remains are also found to the northeast and south of the compound, indicating that the entire compound might have been larger than it appears today.[112] Pottery from the Zebbug Phase has been found, which could indicate that the site was used prior to the construction of the temples, similar to the Skorba site.[113] Also, shards from the post-Temple Period (Borg in-Nadur Phase, 1,500-800 BCE) have

---

109 Stroud, *Prehistoric*, p. 5.
110 Reuben Grima, 'The Landscape Context of Megalithic Architecture', in *Malta Before History*, ed. Daniel Cilia (Malta: Miranda Publisher, 2004), p. 331.
111 Evans, *Antiquities*, pp. 95–103.
112 Stroud, *Prehistoric*, p. 35.
113 Evans, *Antiquities*, p. 102.

been retrieved, suggesting that the Mnajdra site may have had one form or another of usage for between two and three millennia. A special feature about 250 metres behind Mnajdra is the so-called Misqa Tanks, deep cavities carved into the rock that are interconnected by surface grooves. Zammit proposed 'that these tanks furnished water to the Mnajdra and Ħaġar Kim sanctuaries'.[114] Although there is no such evidence, and the origin of these tanks is unknown, Trump supports Zammit's claim.[115]

FIGURE 2.5. Façade of the Mnajdra complex seen from east. Photo: Lomsdalen.

---

114 Zammit, *History*, p. 56.
115 Trump, *Malta: Prehistory*, p. 152; Pace, 'Sites', p. 142.

2.4.3 DESCRIPTION OF THE TEMPLES

THE MNAJDRA EAST TEMPLE

From the modern entrance to the site, the first and smallest temple is the East Temple, a simple trefoil building of Coralline Limestone with what seems to have been an altar in the back niche (Room 10) about 7 metres distant from the entrance, which consists of apparent triple doors leading

into Room 9 (see Fig. 2.4.). This temple has a general southwest orientation towards the Filfla islet and there appears to have been a special relationship between the islet and Mnajdra.[116] That it probably was a sacred islet is indicated by finds of pottery, jars and animal bones dated to the Temple Period; remains probably belonging to a sailor's shrine

---

116 Tilley, *Materiality*, p. 110.

have also been found there.[117] Whether the islet was just visited or settled remains an open question. Richard England further claims that the altar-like shape or bulls' horns profile of the islet attracted the attention of the temple builders.[118]

FIGURE 2.6. East or trefoil temple seen from its south entrance. Photo: Lomsdalen.

117 Stanley Farrugia Randon, *Comino, Filfla and St. Paul's Island* (Malta: P.E.G. Ltd, 2006), p. 43.
118 England, 'Space-Time', p. 413.

The small rubble walls of Mnajdra East are modern reconstructions as the originals have not survived, but the few remaining slabs seem to be original.[119] According to Evans it is difficult to establish an accurate overall plan.[120] Evans further suggests that the restoration gives the impression of a trefoil monument, but that possibly the temple 'originally consisted of two pairs of apsidal rooms, of which the front has entirely disappeared'. According to Evans and Trump, Mnajdra East seems to have been the first and oldest temple, based on its simplicity and the style of pottery retrieved in the soil-pacing in its vicinity, consequently dating it to the Ġgantija Phase (3,600-3,000 BCE).[121] Due to the large quantity of pottery retrieved, Ashby suggests that Mnajdra East was devoted to domestic uses.[122]

THE MNAJDRA MIDDLE TEMPLE

The middle temple is mainly built of the softer Globigerina Limestone and consists of a four-apse structure (Rooms 7 and 8; see Fig. 2.4) with a southeast orientation. It has a central corridor leading to a niche, probably an altar, in the back of the building about 15 metres distant from the temple entrance. In front of the temple is a 7-metre-wide platform elevated above the complex forecourt by about 2.5 metres. It is, according to Trump, 'mostly a modern recon-

---

119 Trump, *Malta: Prehistory*, p. 148.
120 Evans, *Antiquities*, p. 101.
121 Evans, *Antiquities*, p. 103; Trump, *Malta: Prehistory*, p. 148.
122 Ashby *et al.*, *Excavations*, p. 91.

struction', probably from the 1950s, when an extensive remodelling of the front section took place.[123] The present arrangement of the building's front is based on reconstruction work done by Ashby who rebuilt the part of the wall immediately to the northeast and southwest of the entrance with small blocks, as there were no traces of the original orthostats.[124]

Room 7 has a maximal length of about 17 metres and a width of about 8 metres, while the corresponding measurements in Room 8 are 14 and 6 metres. According to Evans, a particular feature of this building is its portal-shaped main entrance, which diverges from the period's usual architectural style.[125] The inside dimensions of the entrance are about 1.6 metres high by 1.25 metres wide and the upper part of the slab is partly broken. Two vertical slabs, a step lower than the entrance, form a passageway into the temple and continue on the other side of the doorway. Albert Mayr found this corridor much damaged; it was later reconstructed by Ashby.[126] To the south of the entrance there seems to have been an auxiliary doorway, not aligned with the central corridor. Room 7, which is open and simple in character, has on each side of the corridor two horizontal slabs supported by vertical orthostats, indicating a demarcated area of special attention: a niche, altar or place to put special objects, a feature also found in the South Temple

---

123 Trump, *Malta: Prehistory*, p. 150; Torpiano, 'Construction', p. 350.
124 Ashby *et al.*, *Excavations*, p. 92.
125 Evans, *Antiquities*, p. 99.
126 Mayr, *Denkmäler*, p. 660; Ashby *et al.*, *Excavations*, p. 92.

Maltese Prehistory: A Literature Review 43

FIGURE 2.7 (above). Broken portal entrance of the middle central corridor as seen from the back altar in Room 8. Photo: Lomsdalen.
FIGURE 2.8 (below). North apse in Room 7; note the harmonious temple with architecture. Photo: Lomsdalen.

(see next section) and in Ġgantija, on Gozo.[127] On the left-hand upright is an engraving of a temple façade, showing four horizontal courses of masonry on top of an orthostat doorway similar to the small slab retrieved from the Mġarr temple site, indicating a roofed temple (Fig. 2.9).[128]

FIGURE 2.9. On the left, the engraved temple façade on the altar wall in Mnajdra Middle Temple. On the right, the Mġarr temple site slab, exposed in The National Museum of Archaeology, Valletta, Malta.[129] Photos: Lomsdalen.

The passage from Room 7 to Room 8 consists of a short paved corridor, threshold slabs and uprights on each side, which widens out into the room, similar to the main entrance and passageway from Rooms 1 and 2 in the South Temple (see Fig. 2.4). According to Evans the original entrance was roofed with lintels, similar to Mnajdra South, which have disappeared.[130] The room has no movable arti-

---

127 Evans, *Antiquities*, p. 98.
128 Evans, *Antiquities*, p. 99.
129 Sharon Sultana, *The National Museum of Archaeology Valletta: The Neolithic Period* (Malta: Heritage Books, 2006), p. 22.
130 Evans, *Antiquities*, p. 99.

cles or equipment and Evans claims there had never been any in antiquity.[131] At the end of Room 8 is a niche, probably an altar, which is aligned along the temple's central corridor line and is, according to Evans, hexagonal in shape (Fig. 2.10).[132] The covering slab has been broken and Evans mentions that Mayr used the broken piece for a different slab, an error duly noticed by Ashby who restored the niche.[133] The covering slab was originally supported by two vertical slabs, between two pillars, but now rests on a modern support. In the southwestern part of Room 8 is another niche with a portal opening about one metre high by about 60 cm wide, flanked by two pillar slabs. At the rear of the room is a table top standing on a circular pillar (similar to Room 3 in Mnajdra South, but less elaborate) between two uprights (Fig. 2.11). In Mayr's time the room was covered with soil, which Ashby cleared.[134] In the northwest wall of the seventh orthostat is a small oval hole, 0.18 by 0.25 metres and 1.30 metres deep, which does not seem to lead anywhere and is of uncertain origin. That the temple was built in the Tarxien Phase is claimed by Evans because a 'fairly large quantity of pottery shards were recovered, all of advanced Tarxien types, which sufficiently established the time of the buildings' construction'.[135] Evans laid out only one trench in the centre of Room 7 of the middle temple and

---

131 Evans, *Antiquities*, p. 99.
132 Evans, *Antiquities*, p. 100.
133 Ashby *et al.*, *Excavations*, p. 93; Evans, *Antiquities*, p. 100.
134 Ashby *et al.*, *Excavations*, p. 93.
135 Evans, *Antiquities*, p. 102.

retrieved only advanced Tarxien pottery.[136] Grima agrees that Mnajdra Middle Temple is more recent than Mnajdra South because its foundation is built against Mnajdra South's outer wall, an argument supported by most archaeologists.[137] Due to the homogeneous construction and architecture throughout the building, Evans further argues that one cannot doubt that it was constructed all at once and not subsequently altered.[138] By this statement Evans seems to attribute an archaic character to the building's core.

THE MNAJDRA SOUTH TEMPLE

The South Temple is a four-apse structure, like the North Temple, but more complex and detailed in its architecture; it shows signs of more than one building phase (Fig. 2.4).[139] The façade has an eastern orientation and is composed of a series of vertical megaliths of Lower Coralline Limestone, six blocks to each side of the main entrance. Besides its concave shape, and the rectangular blocks providing support for its uprights, it bears similarities to the façade of the Ġgantija Temple complex on Gozo, of which the south building is also the most important.[140] Zammit suggests that the horizontal rectangular blocks along the façade have an apparent 'intent to afford seating accommodation

---

136 Evans, *Antiquities*, p. 102.
137 Trump, *Malta: Prehistory*, p. 150.
138 Evans, *Antiquities*, p. 102.
139 Evans, *Antiquities*, p. 102; Pace, 'Sites', p. 129.
140 Evans, *Antiquities*, p. 95.

FIGURE 2.10 (above). Back altar of the Middle Temple. Photo: Lomsdalen.
FIGURE 2.11 (below). Southwest niche of the Middle Temple. Photo: Lomsdalen.

for worshippers'.[141] Whether the forecourt was paved with rough slabs or consisted of solid bedrock is an open question. Ashby found evidence that the forecourt was paved from the façade up to 7 metres out, where the bedrock begins to rise.[142] The threshold slab contains a crystalline vein running along its length, which might have been chosen due to this aesthetic feature.[143] According to my survey of Maltese temples, this is the only threshold that displays such a feature but whether it indicates a special significance of this temple, or just happened to be found in the vicinity is unknown.

A metre and a half into the forecourt from the threshold slab is a hole of 15 cm in diameter, similar to another found in front of the main entrance at Mnajdra's next-door neighbour, Ħaġar Qim. One plausible use of these holes is for tying up animals for ritual sacrifice (Fig. 2.12).[144] The total length of the central corridor is about 15.5 metres from the front to the back wall of a niche (probably an altar) similar to what is observed in every other Maltese temple. On the inside of the entrance uprights are rope holes or the remains of such, which might have been used for swinging doors.[145] This feature is repeated in the entrances to Rooms 2, 3 and 5. Room 1 has a maximal length of about 14 metres and a width of about 7 metres. According to Evans, the floor had probably been covered with *torba*, a plaster-like or ce-

---

141 Zammit, *History*, p. 53.
142 Ashby *et al.*, *Excavations*, p. 94.
143 Stroud, *Prehistoric*, p. 36.
144 Zammit, *Temples*, p. 28.
145 Zammit, *History*, p. 55.

FIGURE 2.12 (above). Rope hole in front of Mnajdra South Entrance. Photo: Lomsdalen.
FIGURE 2.13 (below). Entrance and central corridor of the south temple with its crystalline threshold. Photo: Lomsdalen.

ment material made of limestone, commonly applied to floors and walls in many Maltese temples.[146] The walls, which rise above 4 metres, consist of dressed slabs about 2 metres high and 1.5 metres wide, topped with horizontal elongated blocks, all made of Globigerina Limestone.

In Room 1, especially in the right-hand apse, it is noticeable that whole blocks, not only the exposed faces of the walls, slope inward to narrow the opening of the chamber; this indicates the closing arch of roofing, as is also indicated in the 'Holy of Holies' in the Hypogeum and the small slab model retrieved at the Mġarr Temple site.[147]

On the right-hand side in this apse is a portal entrance with rope holes on the inside, elevated by three steps, leading into Room 5. Room 5 is an L-shaped room consisting of, to the right, another portal entrance into an altar-like chamber and, to the left, an opening of about 0.40 by 0.30 metres into the apse of Room 1, suggested by Trump to be an oracle hole (Fig. 2.14).[148] Skeates suggests that the holes might be later features and seem to have been used for controlled two-way communication.[149] In the northwest corner of Room 5 is another altar-like roofed arrangement which rested on two pillar-like stones at 1.20 metres in height, but is completely collapsed in modern times.[150] According to Pace, Room 5 is from the Tarxien Phase and was fashioned from the wall of Mnajdra South Temple, as orig-

---

146 Evans, *Antiquities*, p. 96; Tilley, *Materiality*, p. 97.
147 Trump, *Archaeological*, p. 29.
148 Trump, *Archaeological*, p. 102.
149 Skeates, *Senses*, p. 170.
150 Evans, *Antiquities*.

FIGURE 2.14. Room 1 in the South Temple with the entrance to Room 5 and the two oracle holes. Photo: Lomsdalen.

inal packing was removed and the former megaliths of the older building were used as a supporting wall of the middle temple.[151] Evans cut two trenches, E and F, in Room 5; in both he found pottery from the Tarxien Phase. However, in trench E, which was cut right inside the portal entrance, he found Ġgantija-style pottery on a lower stratigraphic level—in the thick black earth—than the level where the Tarxien pottery was retrieved.[152] Based on this excavation, Evans concludes that there was a building on this site before the Tarxien Phase.[153] The right-hand apse of Room 1

---

151 Pace, 'Sites', p. 131.
152 Evans, *Antiquities*, p. 102.
153 Evans, *Antiquities*, p. 103.

has also another hole cut in its wall. It is smaller, finer and more recessed than the hole leading into Room 5; it communicates with a chamber that is only accessible from the back side of the temple, which Zammit suggests was used as an oracular chamber.[154]

In the south apse of Room 1 is an altar on the left-hand side with a horizontal slab measuring about 2 metres. Directly opposite is a pitted, hole-decorated portal into Room 3, which, according to Zammit, 'appears to have been peculiarly sacred'.[155] It has similarities to a temple entrance (Fig. 2.15) and is regarded as a modern icon or symbol of the temples; it was chosen to be engraved on the one, two and five cent Maltese Euro coins. On the inside of the doorway are more rope holes. This room consists of two, possibly three, very special double altar-like arrangements, the biggest and most impressive in the front, one to the left and one to the right. The front recessed area or niche is about 2.5 metres wide and has two horizontal slabs. The lower slab is about one metre from the ground, supported by a bi-conical pillar; the top slab, placed about 0.80 metres above the lower one, is supported by a cylindrical pillar. Both pillars are placed at the very centre of the niche. Zammit denominates this room as the 'Sanctuary' and, to quote Trump, 'one would badly like to know what ceremonies took place on

FIGURE 2.15 (right). Above shows entrance to room 3 from room 1 as per today, while below shows the same entrance as it was in 1868. Photo above by Lomsdalen, photo below by the courtesy of National Library, Valletta, Malta.

---

154 Zammit, *Temples*, p. 28.
155 Zammit, *History*, p. 54.

Maltese Prehistory: A Literature Review 53

and around these altars'.[156] According to Evans, the right-hand niche was restored by Ashby who probably erected the table slab that is now supported above the first on a column of 0.75 metres high.[157] Referencing early photos of the site, I question the authentic refurbishment of this specific area. At the height of about 1.9 metres there are, today, clearly visible grooves evidently made to accommodate a lintel or a horizontal slab (as Andrea Pessina and Nicholas C. Vella indicate in Fig. 87), which is no longer *in situ*.[158] The access from Room 3 to Room 2 is hampered by the altar-like arrangement and may originally have been a double altar niche similar to the two niches in Room 3.

At the entrance leading to Room 2 from Room 1 there are, on each side, two altar-like arrangements decorated with drilled holes, similar to what is found in the North Temple. The passage itself is formed of pairs of upright slabs about 2 metres high that widen towards the interior. The uprights are covered with a lintel and the passage floor has a threshold slab with a hole of about 0.18 metres in diameter, which Zammit maintains originally corresponded to a now-missing lintel with a similar drilled hole, providing pivots for a swinging door.[159] According to Evans, Ashby claimed the lintel above the threshold with the drilled hole is a slab now lying on the floor in Room 1, which has a similar hole to the

---

156 Zammit, *Temples*, p. 29; Trump, *Archaeological*, p. 102.
157 Evans, *Antiquities*, pp. 97-8.
158 Luigi M. Ugolini, *Malta: Origins of Mediterranean Civilization*. Eds. Andrea Pessina and Nicholas C. Vella. Trans. Louis Scerri (Malta: Midesea Books, 2012), p. 186.
159 Zammit, *History*, p. 55.

FIGURE 2.16. Above, double altars in Room 3 as seen from Room 2; below, view from Room 3 into Room 2. Photo: Lomsdalen.

one on the threshold.[160] As the lintel was broken and Ashby did not manage to restore it to its original position, he removed it to Room 1 and put it together there, but maintains that the holes must be pivots for a door. Trump suggests another use for this hole: 'Altars and libation holes are common in all temples'.[161] Alternatively, Mario Vassallo maintains that holes in the entrances might have been used to place a pole as a marker for celestial observations.[162] At the back of Room 2 an altar-like niche is centrally placed along the main corridor of the South Temple. The table slab is about 2.7 metres long, 1.25 metres wide and 0.30 metre thick. Mayr found this slab broken, and Evans suggests it was restored by Ashby.[163] The table slab is placed at about 1.05 metres above the ground and is supported on both sides by pillars.

Room 4, situated at the north side of the back room, and a step up from Room 2, is a non-furnished apse. Mayr found this room heavily damaged and filled with debris.[164] Ashby cleared and excavated the room and retrieved from the hard grey earth traces of burnt charcoal, a number of curious clay objects, deformed body parts, pottery, bones, a leg of a clay statuette, other female rough statuettes with pro-

FIGURE 2.17 (right). Above, entrance to Room 2 from Room 1 with altars on each side; below, Room 2's back altar. Photo: Lomsdalen.

160 Evans, *Antiquities*, p. 98.
161 Trump, *Archaeological*, p. 30.
162 Vassallo, 'Sun Worship'.
163 Mayr, *Denkmäler*, p. 659; Evans, *Antiquities*, p. 98.
164 Mayr, *Denkmäler*, p. 659.

truding breasts and swollen body parts and an animal head with the mouth wide open. Zammit suggests that, due to the fact that several of the retrieved figures represent diseased human body parts, the place was sacred to a healing deity.[165] Zammit further claims that the splendid collections of objects obtained from these ruins in 1908, now preserved in the Valletta Museum, 'shows that Mnajdra belongs to the same stage of civilisation as the Hajar Kim and Gozo sanctuaries'.[166]

According to Evans, the pottery retrieved at Mnajdra is mainly from Ashby's excavations, and consists of: 4 shards of Zebbug type, 1 shard of Mġarr type, 34 of Ġgantija type, 700 of Tarxien type, 2 shards of Tarxien Cemetery type and 2 of Borg in-Nadur type. Evans himself, during his 1954 campaign, dug a total of ten trenches 'with varying success, mainly retrieving Tarxien Phase pottery'.[167] However, in two of his stratigraphic cuttings, G and E, he retrieved Ġgantija-type pottery on a lower level than the level containing Tarxien pottery; in both cases he suggests there was a building on these sites from an earlier period.[168] Trench G was cut in front of the threshold entrance to Room 3 and E was dug in Room 5, as previously mentioned.

---

165 Zammit, *History*, p. 55.
166 Zammit, *History*, p. 55.
167 Evans, *Antiquities*, p. 102.
168 Evans, *Antiquities*, p. 103.

FIGURE 2.18. Photo assumed to be from the last decades of the nineteenth century, before major restoration work, showing clear signs of authentic central corridor axes of Mnajdra South. Reproduced with courtesy of the Richard Ellis Archive.[169]

Old photographs of Mnajdra prior to cleaning and restoration do suggest a sense of disorder of Mnajdra South (see Figs. 2.18 and 2.19). Nevertheless, Evans suggests 'the irregularity of the uprights gives the whole façade an archaic look' which is confirmed by pottery evidence; he further points out that the corridor leading into the south building is well preserved.[170] Pace suggests the 'current version

---

169 Richard Ellis, *The Photography Collection: Malta 1862–1930* (Malta: BDL Publishing, 2011), Photo 8, pp. 12–13.
170 Evans, *Antiquities*, p. 96.

FIGURE 2.19. The first known photos ever taken of Mnajdra from 1868, showing disorder, however, indicating an archaic look and original central axes of the temple. Upper left: Mnajdra South entrance seen from southeast. Upper right: Mnajdra Compound seen from northwest. The lower left photo shows the north apse of Room 1, and the lower right photo shows the central corridor as seen from the main entrance of Mnajdra South. All photos appear with courtesy of the National Library of Malta (*Antichità Fenice Nelle Isole Di Malta E Gozo*, 1868).

Maltese Prehistory: A Literature Review

of Lower Mnajdra was built during the Ġgantija Phase'.[171] Evans states that Rooms 2 and 3 of the South Temple may be the oldest (Ġgantija Phase); however, he maintains that Room 1 had 'been made in its present form at the very beginning of the Tarxien phase'.[172]

2.4.4 BUILDING SEQUENCE

The precise dating of the Mnajdra buildings is not free of difficulties, but based on excavation finds and the opinions of the archaeologists cited above, the following chronology for their construction is suggested:

1. Mnajdra East was probably built in the Ġgantija Phase as suggested by Evans;

2. Rooms 2, 3, and 4 of Mnajdra South were built at some point in the Ġgantija Phase (it is unknown whether this occurred before, during or after Mnajdra East was constructed);

3. For the rest of Mnajdra South, the apses of Room 1 may have been completed in the later Ġgantija or early Tarxien Phase and Room 3 may have been refurbished at the same time. (Room 1 is, according to Evans, from the Tarxien period while Pace suggests

---

171 Pace, 'Sites', p. 129.
172 Evans, *Antiquities*, p. 103.

it is Ġgantija Phase construction.) Rooms 5 (Pace suggests it to be Tarxien Phase) and 6 could have been shaped and built in connection with the construction of Mnajdra Middle;

4. Mnajdra Middle was completed sometime into a developed Tarxien Phase.

This chapter has explored the literature reviewing Maltese prehistory within the context of Europe and Mediterranean prehistory. It examined how and why the temples may have come into existence and provided a detailed description of the Mnajdra Temple Complex. The next chapter continues the literature review with a focus on Maltese cosmology and temple period astronomy.

CHAPTER 3

MALTESE COSMOLOGY AND ASTRONOMY: A LITERATURE REVIEW

THIS CHAPTER PROVIDES a literature review of Maltese prehistoric cosmology and archaeoastronomy. It begins by considering evidence for the prehistoric cosmology of the temple builders and describes prior archaeoastronomical studies on the temples.

3.1 COSMOLOGY AND LANDSCAPE

Nicholas Campion defines cosmology as, 'the manner in which human beings relate their cultures to their notions of the nature, order, function, or meaning of the cosmos'; in a Platonic world view, the study of one's self and everything is included in the cosmos.[1] Campion makes it clear that this view contrasts with modern scientific cosmology where the universe is a 'meaning-less' arena in which physical laws work themselves out.[2] Lionel Sims claims that every culture has a 'cosmology', an integrated sum of experiences of their worlds which might include worlds below or above.[3] Reuben Grima defines cosmology as 'the way people understand their universe'; it is about our entire reality, the world in which we live and breathe — not limited to astronomy.[4] To Grima, cosmology is bigger than archaeoastronomy — although it includes that — as it encompasses our entire belief system. Cosmological ideology systematically and repeatedly depicted across various temple sites may suggest that these devices and iconographical illustrations are an integral part of a meaningful order (Fig. 3.1).[5]

Within the temples the different spaces are built differently; a central corridor is often paved with slabs, while the side apses are only paved with *torba*, implying an ideology

---

1 Nicholas Campion, 'Introduction', in *Cosmologies* ed. Nicholas Campion, (Ceredigion, Wales: Sophia Centre Press, 2010), p. 1.

2 Campion, 'Introduction', p. 2.

3 Lionel Sims, 'Coves, Cosmology and Cultural Astronomy', in *Cosmologies*, ed. Nicolas Campion, Proceedings of the Seventh Annual Conference of the Sophia Centre for the Study of Cosmology in Culture, Univerity of Wales, Trinity Saint Dasvid, 6–7 June 2009 (Ceredigion, Wales, UK: Sophia Centre Press, 2010), p. 4.

4 Tore Lomsdalen, 'A Talk with Reuben Grima: Landscape, Cosmology and Iconography Related to Maltese Prehistory', in Appendix II, of this volume.

5 Reuben Grima, 'An Iconography of Insularity: A Cosmological Interpretation of Some Images and Spaces in the Late Neolithic Temples of Malta', *Institute of Archaeology* 12 (2001): p. 55.

Maltese Cosmology and Astronomy: A Literature Review

FIGURE 3.1. Spirals, above, and animal representations, below, from the Tarxien Temples. Photo: Lomsdalen.

of boundaries.⁶ In addition, entrances to special chambers and apses have visible signs of portals, indicating an intention to close off areas as required.⁷ Grima suggests that the spatial order of the temples may represent elements of the builders' world-view.⁸ Low-relief temple carvings of fish, animals and waves may all indicate land and sea, perhaps the most inevitable components of an islander's cosmology.⁹ Grima adds, 'if you are in a coastal context or an island context, the boundary between land and sea becomes a key part of...your understanding of the universe'.¹⁰ Robb goes one step further, maintaining that the Maltese temples stood at the conjunction of two systems of cosmological demarcation, mediating the above-ground living world and the below-ground ancestral world, giving aesthetic associations to the land and the past, linking time and geography.¹¹ The rise of the temples may have involved the formation of new cosmological values connected to the knowledge of geography and landscape, thus composing a new island identity based upon cosmology.¹²

The Maltese archipelago is composed basically of limestone layers of differing hardness, which, through erosion by wind, rain, and sea, created a distinctive, highly seg-

---

6 Lomsdalen, 'A Talk with Reuben Grima', in Appendix II, of this volume.
7 Evans, *Antiquities*, p. 96.
8 Grima, 'Landscape Context', p. 7.
9 Grima, 'Iconography', p. 56.
10 Lomsdalen, 'A Talk with Reuben Grima', in Appendix II, of this volume.
11 Robb, 'Identities', p. 191.
12 Robb, 'Identities', p. 192.

mented landscape of steep slopes, faults, and deep wadis.[13] Grima suggests that prehistoric stone-working activities could only have been undertaken by a society possessing an awareness and knowledge of the availability and quality of the different stones and their suitability for different purposes; their knowledge was likely an important factor in the way they perceived and understood their landscape, resulting in the most remarkable exploitation of stones for monumental building construction.[14] The locations of monumental buildings or temples seem to be based on a meaningful pattern and are placed at the heart of where people might have preferred to live, forming an integral part of a lived-in landscape.[15] Indications of dwellings on pre-temple sites are found in Skorba, and pre-Temple-Period pottery shards have been retrieved at Mnajdra.[16] According to Grima, elevation, slope and surface geology do not seem to have influenced temple locations; instead, proximity to leveled agricultural land, availability of fresh water, south-facing slopes and access to the sea seem to have been strong influencing elements, 'probably because those are the things which matter in choosing a place to live in an agricultural society'; nevertheless, a preference for southern and western locations seems to prevail.[17] Vassallo addressed temple locations cosmologically by observing the relationship between sunrise and sunset positions and

---

13 Grima, 'Landscape Context', p. 327.
14 Grima, 'Landscape Context', p. 328.
15 Lomsdalen, 'A Talk with Reuben Grima', in Appendix II, of this volume.
16 Trump, *Skorba*, p. 10; Evans, *Antiquities*, p. 102.
17 Lomsdalen, 'A Talk with Reuben Grima', in Appendix II, of this volume.

the topography of the horizon. He concluded that at sixteen out of twenty-four temple sites the winter solstice sun rises at the foot of the first hill to the south of the temple; at five others the sun rises at the point where land and sea meet.[18] Vassallo also found that at thirteen sites the sun sets at the foot of a hill; in another three sites it sets where land and sea meet.

3.2 TEMPLE PERIOD ASTRONOMY

The first record of a possible relationship between temples and astronomy came from Vance who published his theories in 1842, especially studying Ħaġar Qim, but also referring to Mnajdra.[19] Vance suggests that the high northeastern vertical pillar at Ħaġar Qim

> was raised for the purpose of tracing with greater accuracy the motions of different planets, and also to contain the libations which were more or less acceptable and efficacious in proportion to the distance at which they were removed from the earth,—the grosser and more impure part of the creation.[20]

Vance further claims that the temple was never roofed as

---

18 Mario Vassallo, 'The Location of the Maltese Neolithic Temple Sites', *The Sunday Times of Malta* (26 August 2007): pp. 44–46.

19 J. G. Vance, 'Description of an Ancient Temple near Crendi, Malta', *Archaeologia* 29 (1842): pp. 231–33.

20 Vance, 'Description', p. 231.

the compound was an ideal spot for worshipping the heavenly bodies and paying 'homage to the sun, moon and stars, to dedicate separate temples to each of the two great luminaries, of a like form and contiguous'.[21] By these statements, Vance not only implies an astronomical but also a cosmological connotation to the temples and further states that decorated slabs next to an altar in Ħaġar Qim were 'designed to symbolize either the sun or the moon, as being the two great causes of nutrition and generation, or the whole globe of the earth in its widest extent'.[22] Vance actually goes one step further in relating the depicted spiral to 'the egg' representing the great mother of the creation in association with the serpent to illustrate the zodiacal circle 'in which the father of all Pagan mythology described his everlasting revolutions'.[23]

Zammit also related the temples to astronomy when, in 1929, he suggested that the pits dug out of a horizontal laying slab at the entrance to the Tarxien Temple represented an image of the stars of Crux (Southern Cross), a constellation clearly visible from Malta in that period.[24] Ugolini also indicates in 1934 a possible relationship between the orientations of the temples and celestial bodies.[25] He also suggests that the Tal-Qadi Stone demonstrates a possible Neolithic 'la lastra astrologica', assumingly meaning a

---

21 Vance, 'Description', pp. 232–33.
22 Vance, 'Description', pp. 233–34.
23 Vance, 'Description', p. 234.
24 Zammit, *Temples*, p. 13.
25 Ugolini, *Origini*, p. 128.

piece, sheet, slab, or a chart of astrology or astronomy.[26] In 1959 Evans found that 'mostly the entrances face in some direction between southeast and southwest', but concludes that apparently orientation was not important and that there was no special interest in heavenly bodies.[27] Trump is in accordance with Evans considering Mnajdra South and goes as far as stating in 2002 that 'astronomical alignments have been suggested'.[28] According to Grima, Trump's reluctance to accept planetary influence regarding the Mnajdra temples has not moderated, and he postulates today that there is 'no reason for us not to accept that this is a characteristic of this building (meaning Mnajdra South) that may possibly be intentional' (see Appendix II).

From then on little or nothing happened on the archaeoastronomy front, until Gerald Formosa discovered and photographically documented summer solstice sunrise and sunset alignments at Ħaġar Qim.[29] Inspired by Professor Thom's research on the megalithic yard, Formosa concluded that the unit of length which Thom proposed was used by the prehistoric megalith builders of Britain and Brittany applies to the Maltese temples too.[30]

Then in 1976 Paul Micallef published two articles which mentioned the orientations of Ħaġar Qim and Mnajdra; however, according to Ventura, he made the wrong inter-

---

26 Ugolini, *Origini*, p. 138.
27 Evans, *Malta*, p. 125.
28 Trump, *Malta: Prehistory*, p. 151.
29 Gerald J. Formosa, *The Megalithic Monuments of Malta* (Vancouver, Canada: Skorba, 1975), pp. 17–21.
30 Formosa, *Megalithic*.

pretation of the precession.[31] In December 1979 Ventura took cardinal measurements of Mnajdra and discovered that Evans' 1971 plan was about 10° off from true North and established its central axis' azimuth of 92.5°.[32] Agius and Ventura then published a booklet in 1980 analysing possible astronomical alignments of the Maltese temples and measuring the central axis orientations of twenty-four temples on Malta and Gozo with a theodolite. Through a chi square test they calculated that the distribution of azimuths based on chance was less than 1 in 1000, and concluded that 'some factor has influenced the choice of orientation of the temples' axes'.[33] They also investigated correlations between temple orientations and the stars, concluding that six temples are oriented towards Sirius and the brightest stars in the constellations of Crux (7,000 years ago) and Centaurus (5,500 years ago).

The next two decades showed considerable interest in astronomical orientations of temples from academics, independent scholars and general enthusiasts. Vassallo has, throughout the 1990s, systematically researched solar alignments of Maltese prehistoric temples; in 2000 he published his findings in three different articles (23 January, 30 January and 6 February 2000) in the *Sunday Times of Malta*, and two articles (6 and 13 February 2011) on a more recent investigation of solar alignments of the Ħaġar Qim Tem-

---

31 Frank Ventura, *L-Astronomija F'malta* (Malta: Pin, 2002), pp. 26–27.
32 Agius and Ventura, *Investigation*, p. 4.
33 Agius and Ventura, *Investigation*, p. 9.

ple.³⁴ Vassallo's investigation focused on whether the temple builders followed a set of common construction rules and put special emphasis on the temples' southwest orientation, their alignments toward the winter solstice sunrise and off-set solar illumination of the left-hand areas within the temples.³⁵ He maintains that the builders wanted to create demarcated areas with clear contrast between light and dark: sunlit main areas, such as the central corridor and the flanking altars, with other areas left in darkness.³⁶ In a later publication Vassallo argues that the Ħaġar Qim Temple was a marker of Neolithic time, based on the yearly movements of the sun.³⁷

In the 1990s Ventura *et al.* pursued their earlier work from the 1970s and 1980s and went from a maximal approach of temple axis orientations used by Agius and Ventura to a minimal one. Thus, the earlier twenty-four orientations were reduced to fourteen.³⁸ Their findings conclude, 'it is clear that they are highly non-random', as they were all within less than a quadrant of arc, from Ġgantija South, with 125.5° to Mnajdra East with 204°, giving a measure of 78.8° of arc (see Fig. 3.2).³⁹

---

34 Vassallo, 'Sun Worship'; Mario Vassallo, 'Ħaġar Qim's Layout Shows Yearly Movements of the Sun', *The Sunday Times of Malta* (6 February 2011); Mario Vassallo, 'Ħaġar Qim: A Leading Marker of Neolithic Time', *The Sunday Times of Malta* (13 February 2011).
35 Vassallo, 'Sun Worship'.
36 Vassallo, 'Sun Worship'.
37 Vassallo, 'Layout'.
38 Foderà Serio *et al.*, 'Orientations', p. 116.
39 Foderà Serio *et al.*, 'Orientations', pp. 116–17.

Maltese Cosmology and Astronomy: A Literature Review

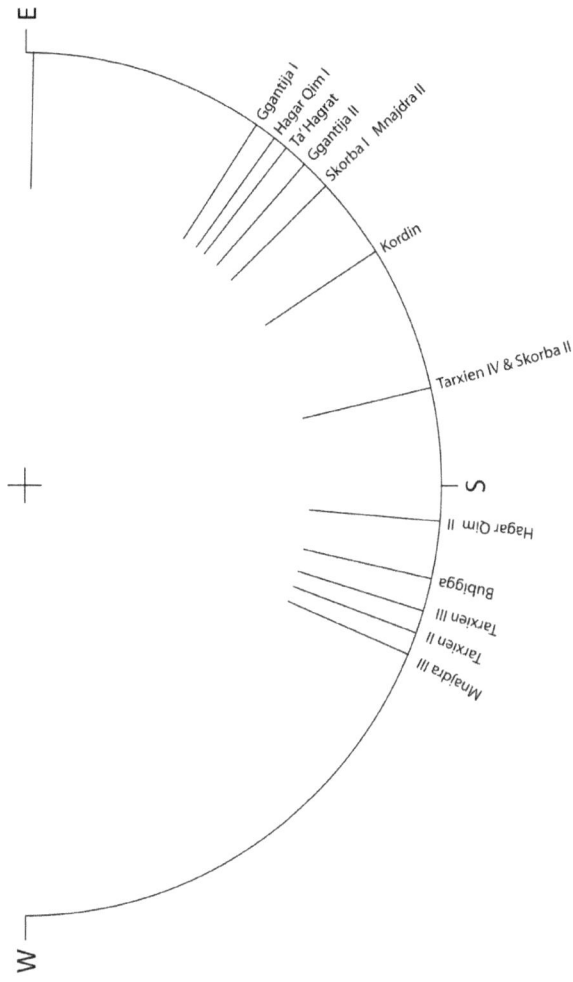

FIGURE 3.2. The orientations of the temple axes.

76 SKY AND PURPOSE IN PREHISTORIC MALTA

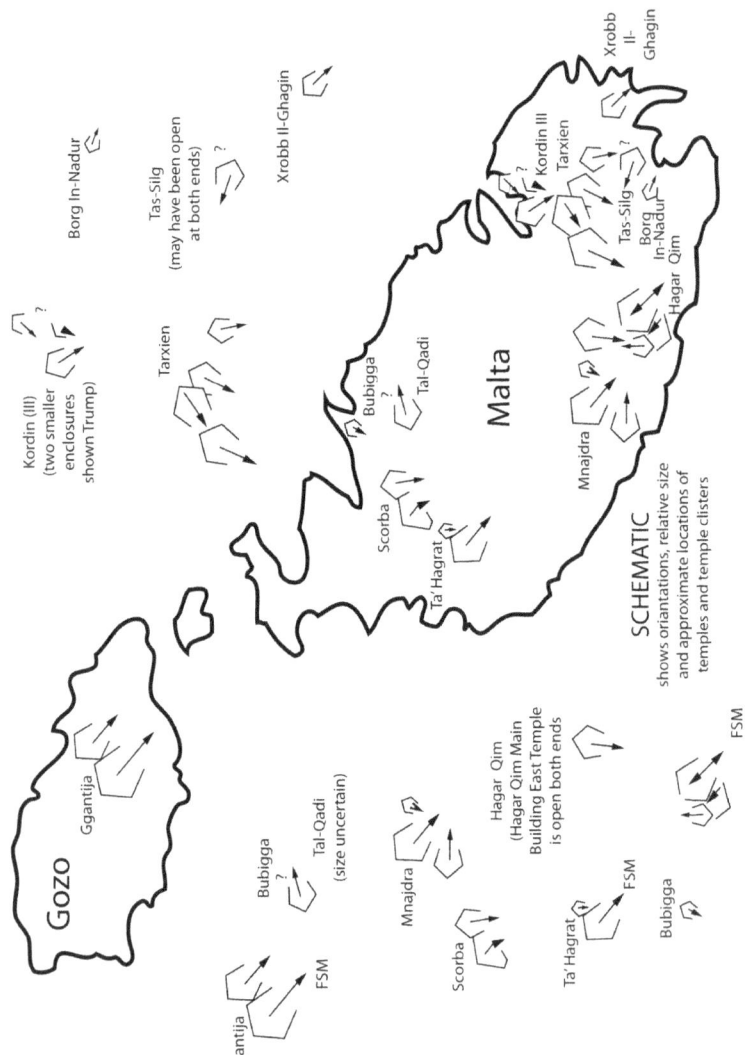

Maltese Cosmology and Astronomy: A Literature Review 77

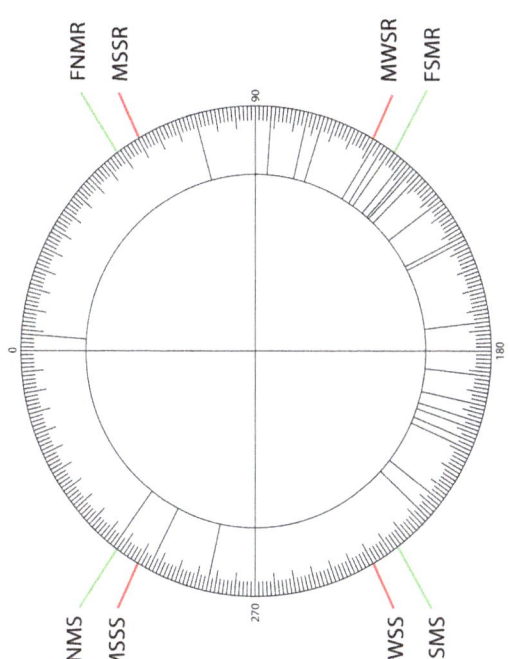

FIGURE 3.3. A schematic showing temple orientations. Above, the temple orientations positioned on the Maltese archipelago. Below, the orientations of 25 temples measured in azimuths looking out of the building. Double entrances at Ħaġar Qim Main Building East and at Tas Silġ are shown as two separate orientations. For the definitions of the acronyms see Appendix IV, Definitions. (Cox and Lomsdalen, 'Prehistoric Cosmology').

In 1999 the second Inspiration of Astronomical Phenomena (INSAP) conference was held in Malta, which seems to have inspired further research in astronomy, archaeoastronomy, phenomenology, cosmology and prehistoric temple culture in general. In 2001, Chris Micallef published an article, which elaborated on a possible reconstruction of the Tal-Qadi stone retrieved from the Ta-Qadi Temple (possibly indicating a crescent moon and stars), showing first quarter, full, last quarter, and new moon intervals.[40] He suggests that the interval of days between these moon phases fits in with the Tal-Qadi stone sequence; however, he does not draw a conclusive result. Further evidence of the temple society's sky interest may be indicated by what is tentatively called a 'solar wheel': a small pottery shard retrieved at nearby Ħaġar Qim.[41] Also in 2001, Klaus Albrecht published his detailed study of temple orientations and examined their possible religious role; he concluded that the builders constructed their temples according to a certain pattern and that it was the alignment to the 'winter solstice which played the decisive role in the construction of the buildings'.[42] Cox also published a paper on temple orientations in 2001, suggesting that they were focused on other astronomical bodies, 'in particular the Moon and southern and far-southern stars'.[43]

A non-astronomical approach to temple orientations

---

40 Chris Micallef, 'The Tal-Qadi Stone: A Moon Calendar or Star Map', *The Oracle, Journal of the Grupp Arkeologiku Malti*, no. 2 (2001): p. 43.
41 Ventura, 'Temple Orientations', p. 312.
42 Albrecht, *Temples*, p. 17.
43 Cox, 'Orientations', p. 36.

is suggested by Stoddard *et al.*, namely a northwest orientation towards Sicily, reflecting ancestral origins and towards Pantelleria, Sicily, and Lipari in terms of the exotic products brought to Malta.[44] Turnbull sustains this argument for temple orientations, claiming that astronomical

FIGURE 3.4. The two tally stones in the trefoil temple directed southwest towards the Filfla Island. Photo: Lomsdalen.

---

44 Stoddart *et al.*, 'Cult', pp. 15–17.

orientations are unlikely given their abstract nature and the difficulty of identifying them other than mathematically; he concludes that orientations towards solstices or major lunar standstills are unconfirmed.[45]

The complex and informative *Malta before History*, written by renowned Maltese scholars and edited by Maltese photographer Daniel Cilia, was published in 2004.[46] In this volume the Maltese architect, Richard England, suggests that the temple builders' interest in cyclic time through sunrise and sunset not only provided a seasonal timing pattern or a marker-system to orient the layout and position of the temple structures, but also represented celestial archetypes, providing the bridge between time as lived by man and cosmic time.[47] England further states, 'the group ritual force which generated the building forms of Ħaġar Qim, Mnajdra and other such sites was born from the belief and conviction that the universe does not function in isolated patterns, but as a whole, totally related to the essence of the COSMOS itself'.[48]

FIGURE 3.5 (right). Above, the eastern and below, the western tally pillars with drilled holes in the trefoil temple. Photo: Lomsdalen.

---

45 David Turnbull, 'Performance and Narrative, Bodies and Movement in the Construction of Places and Objects, Spaces and Knowledges: The Case of the Maltese Megaliths', *Theory, Culture & Society* Vol. 19, no. 125 (2002): pp. 131–32.
46 Daniel Cilia, ed., *Malta before History: The World's Oldest Free-standing Stone Architecture* (Malta: Miranda Publishers, 2004).
47 England, 'Space-Time', p. 412.
48 Formosa, *Megalithic*, p. 12.

Maltese Cosmology and Astronomy: A Literature Review 81

3.3 MNAJDRA AND THE COSMOS

The main features of Mnajdra East and its astronomy are explored in the 1993 work of Ventura *et al.* The small drilled holes on the two vertical entrance pillars to the altar (see Fig. 3.4) were possibly used to tally the days between festivals; the tally on the east and the west pillar which has 179 holes, is close to the number of days in half a year, which then could have been used to sub-divide the year into sequences by repeated observation over longer periods of time.[49]

However, Ventura *et al.* conclude 'the holes are a tally', probably of days of a regular and significant sequence of events that occurred annually and further that 'these events included the heliacal rise of the Pleiades of other stars and asterisms; but we are some way from proving this'.[50] Ventura *et al.* suggest that holes may actually be a Maltese form of parapegma, which is, according to Neugebauer, 'a style of calendar based on the rising or settings of stars'.[51]

In addition to the rows of holes on the east pillar there is, further down, a different group of six smaller holes (Fig. 3.4, barely visible lower part centre), arranged in a roughly circular pattern, which Ventura *et al.* suggest might represent the Pleiades. According to their work this sequence

---

49 Frank Ventura *et al.*, 'Possible Tally Stones at Mnajdra, Malta', *JHA* 24 (1993): p. 178.

50 For definition of 'heliacal rise', see Appendix IV. Ventura *et al.*, 'Tally', p. 182.

51 Otto Neugebauer, 'An Arabic Version of Ptolemy's Parapegma from the 'Phaseis'', *Journal of the American Oriental Society* 91, no. 4 (1971): p. 506.

FIGURE 3.6. WSSR through the broken portal entrance of the middle temple.
Photo: Lomsdalen.

started with the heliacal rising of the Pleiades, which Dicks states 'has been used by various people all over the world to mark the passage of time and the seasons of the year'.[52] Anthony Aveni maintains that the Pleiades have been univer-

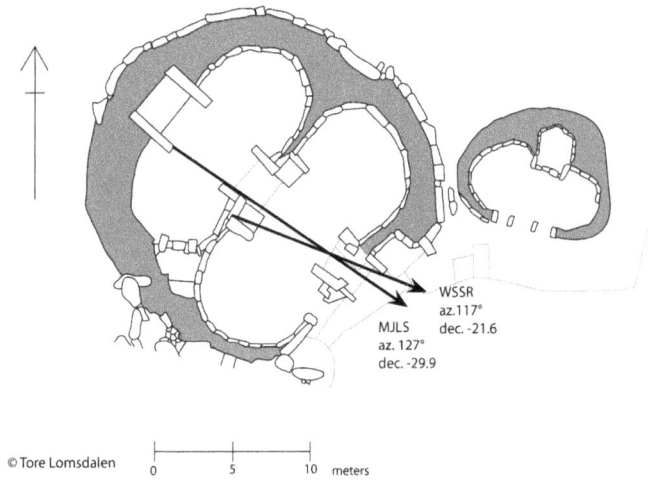

FIGURE 3.7. Constructed plan of WSSR and MJLS alignments at the middle temple.

sally recognised in a variety of ancient cultures.[53] Ventura *et al.* also note that later Egyptian culture monitored subsequent heliacal risings for time-keeping, and conclude that

---

52 Ventura *et al.*, 'Tally', p. 180; D. R. Dicks, *Early Greek Astronomy to Aristotle* (Ithaca, NY: Cornell University Press, 1970), p. 10.
53 Anthony Aveni, *People and the Sky: Our Ancestors and the Cosmos* (London: Thames & Hudson, 2008), p. 10.

in prehistoric Malta this might also have been plausible.[54] Michael Hoskin also suggests that the Pleiades rose with a declination of 0° east during the Temple Period in Malta and had significance for early peoples both in the Old and the New World.[55] Because of this, Ventura and collaborators have suggested the easterly alignment of the Mnajdra South Temple might have marked the rising of the Pleiades, not the equinoctial sunrise.[56]

Few studies regarding Mnajdra Middle Temple and its possible celestial alignments have been done. Vassallo observes that the winter solstice sunrise (WSSR) illuminates 'the walls of the apses behind the altars'; a similar observation is made by Albrecht (Fig. 3.6).[57] In private correspondence with Cox in 2010, speculations that the altar in Room 7 of the Mnajdra Middle Temple might have a major lunar standstill alignment (MJLS) were discussed (see Fig. 3.7).[58] Easy observation of the far-southerly moonrise is limited to a season: January to June.[59] The last time direct observation of a major lunar standstill was possible occurred from May 2005 to June 2007; the next occurrence will be visible around the year 2026.

In the 1970s and 1980s astronomical studies related to

---

54 Ventura *et al.*, 'Tally', p. 170.
55 Michael Hoskin, *Tombs, Temples and Their Orientations: A New Perspective on Mediterranean Prehistory* (Cambridge: Ocarina Books Ltd., 2001), p. 31.
56 Foderà Serio *et al.*, 'Orientations', pp. 116–17.
57 Vassallo, 'Sun Worship', pp. 36–37; Albrecht, *Temples*, p. 25.
58 See Appendix IV for definition of MJLS. John Cox, Correspondence, 4 February 2010.
59 Cox, 'Observations of Far-Southerly Moonrise', p. 347.

the Mnajdra South temple came mainly from three sources: Agius and Ventura, Paul Micallef, assisted by Alfred Xuereb, and the team of Ventura, Hoskin, and Foderà Serio. Micallef mainly studied sunrise illumination at the equinoxes and solstices and says, 'the slit images produced at the Mnajdra Temple qualify the monument to be seriously considered as a solar calendar or observatory'.[60] Micallef's conclusion indicates clearly that the Mnajdra South Tem-

FIGURE 3.8. Slit image of the illumination of the south temple at solstice and equinox. Photo: Lomsdalen.

---

60 Micallef, *Calendar*, p. 26.

ple is the only solar temple in the Maltese islands.[61] However, Agius and Ventura questioned whether the temple was purposely constructed for solar alignments at the equinoxes and solstices; they claim it would be nearly impossible for the builders to find the equinoctial point without first determining the solstitial sunrise positions, then bisecting the angle between the two positions.[62] Based on this argument, Ventura and Agius organised a search in 1981 for possible postholes along the direction lines of the winter and the summer solstices as seen from Mnajdra South. Two corresponding human-made holes were found in the horizon near Mnajdra. The hole for summer solstice sunrise is not perfectly aligned with the Mnajdra South Temple as the sun rises about 3° further north than its position; however, the posthole for the winter solstice sunrise is perfectly aligned with the Mnajdra South Temple (see chapter 5.2.2).[63]

In the 1990s, Maelee Thomson Foster was the first to investigate further divisions of the year as defined by the structure of the South Temple.[64] Later independent research regarding this topic was done by Chris Micallef.[65] Thomson Foster suggests that the sunrise illuminates de-

---

61 Micallef, *Calendar*, p. 41.
62 Agius and Ventura, *Investigation*, p. 18.
63 Ventura *et al.*, 'Tally', pp. 173–75.
64 Maelee Thomson Foster, 'Orientation: A Design Determinant Is Utilized as a Means to Explore the Maltese Lower Temple of Mnajdra as a Possible Solar Calendar' (paper presented at the INSAP II, International Conference on the Influence of Astronomical Phenomena Malta, 7–14 January 1999).
65 Chris Micallef, 'Alignments Along the Main Axes at Mnajdra Temples', *Journal of the Malta Chamber of Scientists* 5, nos. 1 and 2 (2000).

marcated areas within the temple at the so-called cross-quarter days (halfway between the equinoxes and solstices) and remarks that these days marked four major festivals in later Celtic culture. Micallef went one step further, suggesting that solar illumination of specific areas on the

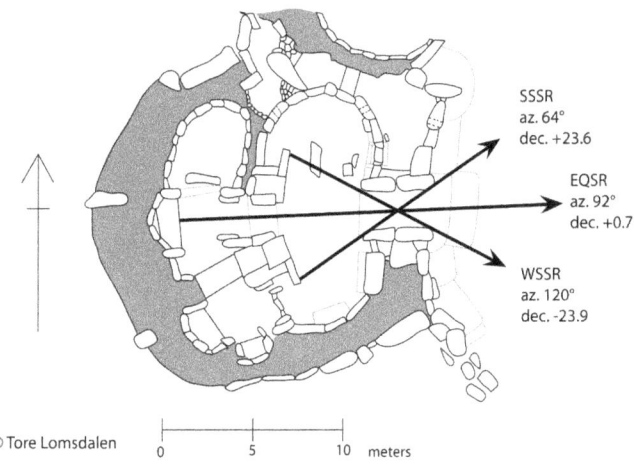

FIGURE 3.9. Constructed plan of equinoctial and solstitial illumination of the south temple.

so-called eighth-days (days between either an equinox/solstice and a cross-quarter day) can also be seen.[66] Micallef seems more determined in his conclusion, declaring that the temple is an 'astronomical observatory since the images hit important stones inside the solar temple throughout

---

66 Micallef, 'Alignments', p. 10.

the year'.[67] Both Foster's and Micallef's work investigates whether the temple builders intended to follow the arc of the sun throughout a full year, not only at the equinoxes and solstices.[68]

This chapter has reviewed the literature on cosmology and astronomy in the context of the Maltese temple culture. It describes in detail the evolution of astronomical observations and their importance for the Mnajdra Temple complex. The next chapter will explain the methodology employed in this work.

---

67 Micallef, 'Alignments', p. 15.
68 Thomson Foster, 'Orientation'; Micallef, 'Alignments'.

CHAPTER 4

METHODOLOGY

THE RESEARCH METHODOLOGY of this project consisted of fieldwork, with the active research conducted in multiple site visits to both the Mnajdra temple complex and other Maltese prehistoric temples. Site visit activities consisted of surveying, astronomical observation and photography, phenomenology and experimental archaeoastronomy. Each of these aspects of the site visit work is described separately in this chapter.

SITE VISITS

Site visits and fieldwork started in 2010, with the observation of the horizon point of the spring equinoctial sunrise

(EQSR) in Mnajdra. I made a total of eleven visits to the Maltese archipelago from that date until January 2013. A visit lasted approximately one week, giving enough time to adapt the observations to unforeseen changes in atmospheric conditions. The total number of visits was based on evolving research developments. For observing equinoctial and solstitial sunrises under perfect weather conditions a total of three visits to Malta might have been sufficient. However, an in-depth study which included prehistory raised additional questions. Establishing personal contact with Maltese scholars (Grima, Vassallo, Ventura) who have studied the prehistoric temples in the last decades gave valuable input, which extended my research objectives. Conversing with and interviewing the above-mentioned persons became part of my research data. At Mnajdra South when the sun rises at the winter and summer solstices, the observation time is brief due to the narrowness of the visual alignment towards the sun (see Figs. 5.7 and 5.8); thus, it was difficult to foresee all factors involved in the events.

SURVEYING

Although this research was primarily dedicated to the Mnajdra complex, other prehistoric sites in Malta and Gozo were also visited for two purposes: firstly, to establish any possible structural and architectural coherence of the prehistoric temples and sites throughout the archipelago, and secondly, to search for confirmation of consistency of orientations and astronomical alignments. The temples visited include: Ħaġar Qim, Ġgantija, Xagħra Circle, Tarxien, Hypogeum, Kordin III, Ta' Ħaġrat, Skorba, Tal-Qadi, and Bugibba.

The above-mentioned temples' main axes and crossjambs were measured at each site. This included obtaining the horizon altitude for these directions.[1] Azimuths and altitudes were measured using a handheld Suunto tandem (compass and clinometer) with a 1° precision. Magnetic anomaly was verified locally; however it was found negligible when compared with compass readings for the azimuth of the central axis of orientation of the South Temple done by Ventura *et al.*, who used a theodolite.[2] All magnetic azimuths were then adjusted to true azimuth using a magnetic declination of +2° 15' for Malta in the period 2010–2013, obtained from the Natural Resources Canada website.[3] A Garmin 12 handheld GPS was used to establish location; when averaged over time, this GPS can reach a precision of 5 metres. In addition to these tools, a 30-metre tape was used to take site measurements and determine mid-points of entrances and corridors.

The main axes were established by erecting a portable pole at the centre of the main entrance threshold and positioning myself against the back of the entrance. This allowed the compass bearing to be measured. The same

---

1 For definition of 'horizon altitude', see Appendix IV.
2 Ventura *et al.*, 'Tally', p. 172.
3 Natural Resources Canada, 'Magnetic Declination Calculator', Federal Identity Program, http://geomag.nrcan.gc.ca/calc/mdcal-eng.php [Accessed 10 June 2010].

procedure was employed for all measured orientations of the temple's central axes as well as all alignments towards celestial bodies.

ASTRONOMICAL OBSERVATION AND PHOTOGRAPHY

Photographic documentation played a major role in this field research project. Malville stresses the importance of 'observations' in archaeoastronomy.[4] Based on his own field experience, Malville urges that all survey measurements should be complemented by photographic evidence whenever possible. Photography provides visual, factual evidence of specific events in time as a reference for future documentation and dispute, by illustrating both the actual sunrise at selected points on the apparent horizon and the illumination of sunrays onto demarcated areas within the temples.

Obtaining pictures of equinoctial, solstitial, crossquarter and eighth-day sunrises (when present in person), involved sitting or kneeling with my camera on a tripod

FIGURE 4.1 (right). Above, sitting in front of the back altar and photographing the EQSR; below, photographing the SSSR from the vertical orthostat in the south temple. Photo: Lomsdalen and Silva.

---

4 J. MacKim Malville, 'Archaeoastronomy', Lecture given at the MA course in Cultural Astronomy and Astrology at the University of Wales Trinity Saint David, 28 February 2010.

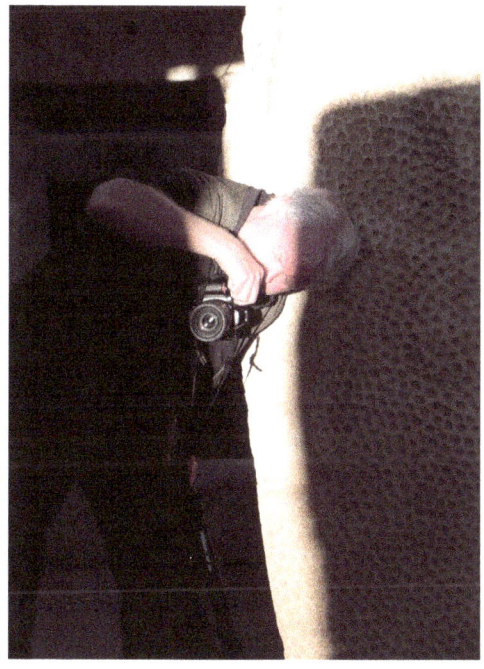

in front of the pre-determined spot where the expected illumination would be cast onto the temple wall and the sun when it started to rise on the apparent horizon. In this case there are actually triple lines as the illumination is the backline, the temple entrance is the middle line and the sunrise is the frontline, helping to establish the correct coordinates. As I worked alone, I sometimes used two cameras in order to take adequate photos of both the sunrise and the illumination onto the temple wall. A single session could produce several hundred photos, allowing the most relevant to be selected and edited later. My visit to Mnajdra South for the winter solstice in 2012 was directed towards photographic documentation of the sunrise illumination of the oracle hole in Room 5 through a hole in the temple façade (Fig. 5.14). To accomplish this I sat beside the oracle hole to observe and photograph the sunrise through the hole in the temple facade over the winter solstice sunrise (WSSR) posthole where I had previously placed a pole, again a triple coordinate line-up example (Fig. 5.13).

Part of my preparation involved pinpointing the equinoctial and solstitial sunrise points as well as the theoretical rising point of the most southerly moonrise seen from Mnajdra Middle before actually observing the events. For these events I had collected astronomical and coordinate information regarding when and where relevant celestial bodies would rise on the apparent horizon, and aligned myself inside the temples accordingly to document these events photographically. The following section describes the techniques used in more detail.

Visits were primarily scheduled around the equinoxes

and solstices. However, in order to fully investigate the theories in Thomson Foster's and Micallef's studies, additional visits to Mnajdra were conducted in 2011 and 2012 to observe the effects of solar illumination around other clearly defined periods of the solar year: the cross-quarter days and eighth-days.[5] For observation of the eighth-days, the site was visited on 7 November 2011 and on 26 February 2012; observations at a cross-quarter day were made on 6 May 2012. With the help of an on-site photographic assistant, all cross-quarter days and eighth-days from winter solstice (2011) to summer solstice (2012) were photographically documented.

EXPERIMENTAL ARCHAEOLOGY

Archaeology employs the well-known principle of testing a hypothesis through experiment. According to Mathieu, experimental archaeology aims, within the context of a 'controllable imitative experiment, to replicate past phenomena in order to generate and test hypotheses to provide or enhance analogies for archaeological interpretation'.[6] Renfrew and Bahn illustrate this aim with the experimental 1960 earthwork constructed on Overton Down, southern England, in order to study its formation processes through

---

5 Thomson Foster, 'Orientation', p. 12; Micallef, 'Alignments'.
6 James R. Mathieu, ed., *Experimental Archaeology, Replicating Past Objects, Behaviors and Processes* (Oxford: Archaeopress, 2002), p. 12.

time.[7] Another example was the failed attempt, in 2000, to transport a three-ton bluestone, like those used in Stonehenge, from its source in West Wales to the Salisbury Plain, using only the available technology of the time.[8] Another example was the 1947 Kon-Tiki balsa raft, built by Thor Høyerdahl to sail from Peru to Polynesia, in order to demonstrate that cultural exchange between South America and the Polynesian islands was possible.[9]

In archaeoastronomy this approach seems to be less implemented. An example of such an approach is given by Malville who suggested that the moon at its major lunar standstill could be seen to rise between the twin pillars of Chimney Rock in Chaco Canyon (Colorado, USA), and that 'this location might contain evidence of ancient astronomy practiced by its inhabitants'.[10] This phenomenon was photographically validated by the observation of said moonrise on 8 August 1988 and again in the autumn of 2004.[11]

Analogous to Malville's investigation, I performed a similar test using Jupiter's rise at declination 0° on 23:19 UT on 26 June 2010 (Fig. 4.2). Since it was known that Jupiter would rise at the same location on the horizon that the sun would rise on the equinoxes, I sought visual confirmation

---

7 Colin Renfrew and Bahn Paul, *Archaeology: Theories, Methods and Practice* (London: Thames & Hudson, 2008), p. 55.

8 BBC News, http://news.bbc.co.uk/2/hi/uk_news/wales/731519.stm [Accessed 11 March 2013]

9 Thor Høyerdahl, *Kontiki Ekspedisjonen* (Oslo: Gyldendal Norske Forlag, 1948).

10 Malville, *Prehistoric*, p. 88.

11 Malville, *Prehistoric*, pp. 88–90.

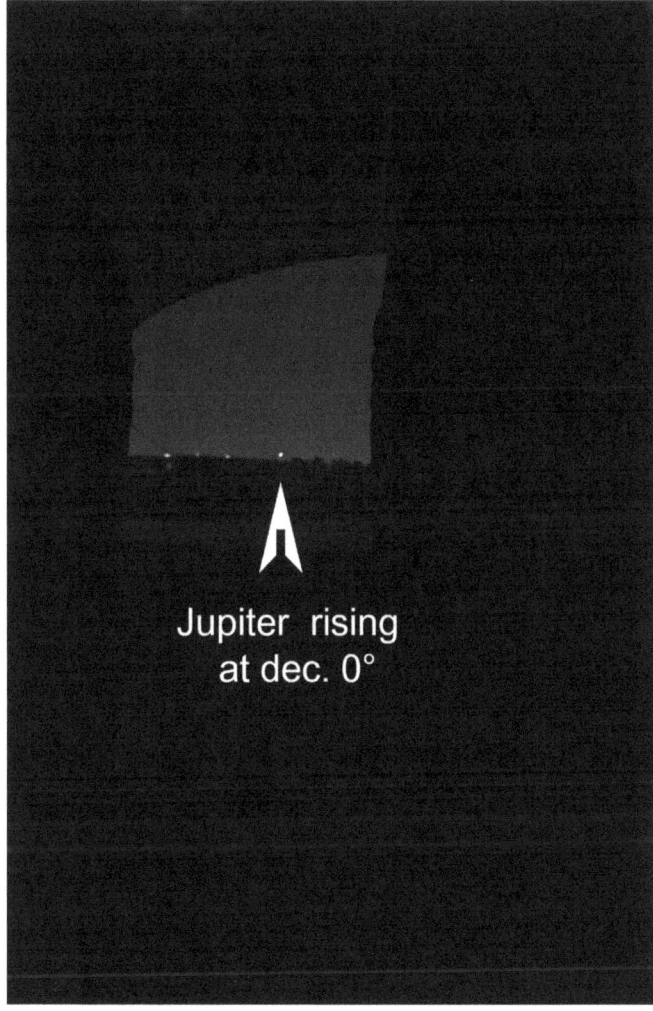

FIGURE 4.2. Jupiter rising on 26 June 2010 at 23:19 UT. Photo: Lomsdalen.

of the location of this point and its relationship with the central axis of Mnajdra South (Fig. 4.2); the sun is too bright to allow for accurate positioning with a clear eye as it rises. My experiment therefore used the rise of Jupiter at declination 0° instead of the sunrise at equinox.

The previously remarked discovery of human-made postholes on the horizon of Mnajdra, which were closely positioned to the WSSR and summer solstitial sunrise (SSSR), provided the perfect scenario to implement a small experiment. Although the hypothesis that these postholes are connected to Mnajdra South was tested by Ventura *et al*. in the early 1980s, I decided to do my own investigation of Ventura's results. My investigation involved placing posts in both the SSSR and WSSR postholes.

For the SSSR hole a 5-metre-long bamboo stake with a white cloth at the top was used, while on the WSSR hole a piece of wood about 1.5 metres tall with a white piece of cloth in front was used to aid visibility from Mnajdra South. When first observing the alignment of the WSSR hole in 2010, a member of the on-site staff stood precisely on the hole with a white cloth in front of him. This important and unique experiment took place on 26 June 2010 at 18:38 UT to see if the moon would rise with the same declination as did the sun during the Temple Period. This was done in hopes of strengthening the hypothesis that the posthole and the WSSR alignment could have been intentionally related, even though the dating of this hole has yet to be defined (see chapter 5.2.2 and Fig. 5.10).

The time and date for the experiment of the WSSR over the horizon postholes was conducted first on 28 December

FIGURE 4.3. Portable poles used for front and back sight alignments. Left, the view from Mnajdra Middle towards the WSSR posthole; right, the same alignment seen from the entrance to Mnajdra Middle. Photo: Lomsdalen.

2010 at 06:24 UT without a clear position of the sunrise in relation the posthole; however, it was successfully accomplished on 17 December 2011, at 06:10 UT. A similar experiment of the SSSR was achieved on 24 June 2011 at 04:22 UT.

PHENOMENOLOGY

This research is based on a primarily quantitative methodology, but a certain qualitative approach aided analysis of research data and the search for the meaning and astronomical intentionality behind the construction of the Mnajdra temples, namely, by walking and experiencing the landscape. According to Tilley, 'phenomenology involves the understanding of description of things as they are experienced by a subject'.[12] It is about the relationship between Being and Being-in-the-world and creating a gap, a distance in space, as people are setting themselves apart from an object.[13]

Thomas refers to 'monuments as an undifferentiated class of phenomena' since a given site could function for various purposes at different times throughout history.[14] Ruggles employs a sensory or phenomenological approach when he tries to understand prehistoric people's patterns of perception in the relationships between landscape and

---

12 Tilley, *Phenomenology*, p. 12.
13 Tilley, *Phenomenology*, p. 12.
14 Thomas, *Understanding the Neolithic*, p. 224.

their monuments.[15] He concludes that we can actually never know how the prehistoric observer would experience such relationships; nevertheless, the attempt deserves serious consideration. The challenge is to do so without bringing personal viewpoints and modern judgments into the context of Maltese prehistoric temple phenomenology.[16]

Although my fieldwork was not initially designed with phenomenology in mind, visiting prehistoric temples at specified times created a certain sensory experience, and a relationship with the 'spirit of place' could be perceived, similar to what has been suggested by Graves and Poraj-Wilezynska.[17] Being in the presence of the majestic atmosphere of such a dramatic site as Mnajdra, spending nights in solitude, hearing cracking noises, with only the starry sky, earth, and sea for company can inspire metaphorical, even testable, suggestions about the location and usage of the temple. This approach actually inspired my archaeoastronomical experiments described in the previous section. Experiencing the dramatic sunrises at the equinoxes and solstices from inside the temple inspired reflections about the purpose and understanding of what kinds of activities might have been conducted on the site. An ever-present

---

15 Clive Ruggles, *Astronomy in Prehistoric Britain and Ireland* (London: Yale University Press, 1999), p. 151.

16 Charlotte Aull Davis, *Reflexive Ethnography: A Guide to Researching Selves and Others* (London: Routledge, 2008), p. 7.

17 Tom Graves and Liz Poraj-Wilezynska, '"Spirit of Place" as Process: Archaeography, Dowsing and Perceptual Mapping at Belas Knap', *Time and Mind: the Journal of Archaeology, Consciousness and Culture* 2, no. 2 (2009): p. 185.

sense that one is not alone might relate to Skeates' suggestion that 'spatiality' is relevant in the Maltese pre-historical context in reference to the bodily movements of persons through monuments and landscapes, particularly during ritual performances.[18] Being and walking in the landscape, trying to perceive how the prehistoric people experienced living in and employing the surroundings was done in hopes of gaining such insights, inspired by Merleau-Ponty's argument that 'phenomenology is the study of essences', and brings in the examples of essence of perception and essence of consciousness, but also puts essences back into existence and 'offers an account of space, time and the world as we "live" them'.[19] Phenomenology is not solely the reflection of a pre-existing truth, but, like art, the act of bringing truth into being.[20]

This chapter elucidates the methodology implemented for achieving the aim and objectives for my field research. It describes the methods used during field surveys and visits and details the use of photography for archaeoastronomical documentation. The chapter further describes the use of experimental archaeoastronomy to verify modern assumptions of temple orientation and alignments. Finally, it examines phenomenological components of my fieldwork. The next chapter will present the results of my fieldwork.

18 Skeates, *Senses*, p. 68.
19 Merleau-Ponty, *Phenomenology*, p. vii.
20 Merleau-Ponty, *Phenomenology*, p. xxiii.

CHAPTER 5

RESULTS

THIS CHAPTER PRESENTS the results of my fieldwork in Malta, with emphasis on the Mnajdra Temple Complex. Details of the findings, which include survey measurements and documented observations of celestial events, are presented and the results compared to other researchers' findings where appropriate.

Table 5.1 summarises fieldwork measurements compared with previous research on azimuths (Az), altitudes (Alt), and declinations (Dec) of the Maltese Temples' central axes. My own measurements are given in the right hand column, listed under Lomsdalen. The table shows that my own measurements of Maltese Prehistoric Temples coincide with those of previous researchers, ensuring that my fieldwork can be applicable for comparisons to other relevant studies and in particular to Mnajdra.

Temple	Notes	A & V 1981 AZ	A & V 1981 DEC	F, H & V 1992 AZ	F, H & V 1992 ALT	F, H & V 1992 DEC	COX 2001 AZ	ALBRECHT 2007 AZ	LOMSDALEN 2010-13 AZ	LOMSDALEN 2010-13 ALT	LOMSDALEN 2010-13 DEC
Ggantija S		128	-29.9	125.5	1	-27.3	129	128	128	1	-29.2
Ggantija N		133	-33.5	134.5	1	-33.8	135	132	137	1	-35.5
Hagar Qim North Temple		186	-53.7	186.0	-0.4	-54.1		180	184	0	-54.0
Hagar Qim Main Temple	CORRIDOR	128.7	-30.5	129	-0.4	-31	131	128	128	1	-29.3
	ROOM 13	(213)	-42.8				214	205	212	0	-43.4
	ROOM 12	(255)	-12.1				255	254	258	0	-9.7
	ROOM 11	0.8	54.4				4	359	5	0	53.9
Mnajdra South		92.7	0	92.7	3.72	0	94	92	92	4	0.7
Mnajdra Middle		138.1	-38.1	138.5	-0.3	-37.6	139	138	140	-1	-39.1
Mnajdra East		207.2	-46.1	204	-0.3	-48	207	207	200	-1	-50.5
Skorba E (II)		168.5	-50.1	168.5	2.5	-50.1		168	172	B	
Skorba W (I)		134.9	-33.3	138	2.28	-35.3		134	137	B	
Ta'Hagrat I		130.6	-29.4	131	3.37	-29.7		130	131	3	-30.0
Tarxien East Temple		(170)	-52.9	168.5	0.6	-52			182	B	
Tarxien Main Temple	S(I)	200.2	-49.5	201	0.7	-48.5		205	204.5	B	
	N(II)	230.1	-31.3					229	227	B	
	N FORECOURT	142	-39.7						139	B	
	E (III)	198.7	-50.1	198	0.7	-49.8		202	202	B	

TABLE 5.1. Summary of fieldwork measurements compared with previous research. 'Alt'. indicates the horizon altitude in degrees. Abbreviations: **A & V 1981**: Agius and Ventura, 'Investigation', p. 13. **F, H & V 1992**: Foderà Serio *et al.*, 'Orientations', p. 115. **COX 2001**: Cox, 'Orientations', p. 28. **ALBRECHT 2007**: Albrecht, *Temples*. **LOMSDALEN**: Present field survey. **B**: blocked by modern buildings

5.1 ARCHAEOASTRONOMICAL SURVEY OF MNAJDRA EAST AND MIDDLE TEMPLES

Each temple within the Mnajdra complex has its own unique orientation. Mnajdra East is oriented towards south/west and Mnajdra Middle towards the southeast. Mnajdra South, on the other hand, has an eastern orientation, the only extant temple on Malta with such a placement (Fig. 2.4).[1]

5.1.1 MNAJDRA EAST

The East Temple is oriented largely towards the south and directed towards the Filfla islet situated about five kilometres off the coast of Mnajdra. Tilley suggests that the temple entrance looks directly towards the island and therefore might have a special relationship to it.[2] It faces an important local landscape feature comparable to the Naffara Hill, which the Ġgantija temple faces. The East Temple's central axis has an azimuth of about 210°, directly towards Filfla. The islet's plateau, as seen from the temple, presents a horizon altitude of -1° (declination -45.42°).

---

[1] Author's note: However, after concluding the present research program, the author has investigated another three sites in March 2014 with an indicative eastern orientation. One is the Tas-Silġ in Malta and the two others are Ta' Marziena and the Xagħra Circle in Gozo.
[2] Tilley, *Materiality*, p. 110.

108 SKY AND PURPOSE IN PREHISTORIC MALTA

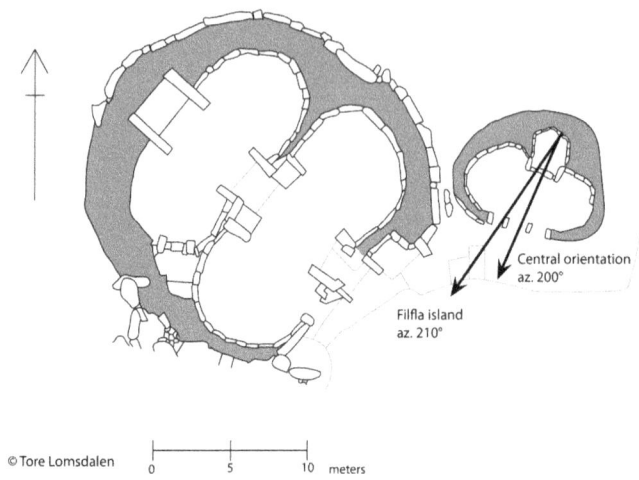

FIGURE 5.1. East temple with its central orientation at az. 200° and Filfla Island at az. 210° seen from the back niche.

5.1.2 MNAJDRA MIDDLE

The observations made by Vassallo and Albrecht were confirmed in 2010 and 2011, where the first rays of the WSSR were observed from the left (south) side of the altar. The sun was seen to rise at an azimuth of 117° in a cross-jamb view through the portal entrance. In the current century the first rays of WSSR cast a very demarcated illumination onto the left-hand side altar; an hour or so later, a general illumination is cast into the temple until the sun disappears behind the protective tent put up in 2009.

Since repeated site visits called into question the Middle

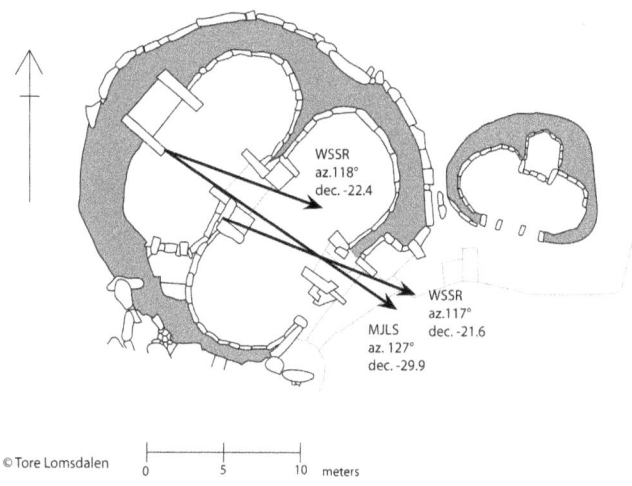

FIGURE 5.2. Middle temple with alignments towards major lunar standstill (MJLS) and WSSR altars in Room 7 and 8.

Temple's lack of any astronomical alignments other than those observed by Vassallo and Albrecht, a more thorough investigation was conducted. A portable pole placed in the middle of the portal entrance was measured from the centre of the south altar slab at the end of the corridor of Room 7, which is illuminated on WSSR. Another pole was set at the WSSR posthole on the horizon, identified by Agius and Ventura in 1980.[3] Observing from the centre of the altar slab, the pole in the centre of the portal entrance is perfectly aligned to the posthole with an azimuth of 120°

---

3 Lomsdalen, 'A Talk with Frank Ventura', in Appendix I, of this volume.

(declination -24.55°). Unlike the South Temple, for which this posthole is a perfect match to the WSSR orientation (see below), the North Temple is about three metres higher than the South Temple, reducing the horizon altitude at the posthole to -1°, thus giving the quoted declination. However, as seen from the North Temple, the posthole clearly marks the location where land and sea meet at the horizon.

During the 2011 WSSR it could be observed that, from the above-mentioned altar in the North Temple through the middle of the portal doorway, the sun rose about six sun-disks further north than the posthole location, at an azimuth of 117° and a horizon altitude of +0.5°(declination -21.60°; see Fig. 5.3).

At the south end of Room 8 in the North Temple is a niche or altar with a portal entrance. The azimuth from the centre of this altar through the portal entrance is 0° (true north) which agrees with Albrecht's findings.[4] On the opposite apse wall of the altar is a hole apparently leading nowhere with an azimuthal orientation of about 210° when seen from the inside out.

From the far back altar the central axis' azimuthal orientation is 140°. From the left (south) edge of this altar, two orientations were measured. One is a diagonal or cross-jamb view with an azimuth of about 118° (declination about -22.37°). If the front apses were not yet constructed, this area would have received offset illumination at the WSSR around 2,500 BCE—late Tarxien Phase—when this temple

---

4 Klaus Albrecht, *Maltas Tempel: Zwischen Religion und Astronomie* (Wilhelmshorst, Germany: Sven Näther, 2004), p. 56.

FIGURE 5.3. WSSR seen from the corner of the left altar in Room 7 of the middle temple; a marker centrally placed in the temple entrance and a post on the WSSR horizon hole (az. 120°) are clearly visible. Note the sunrise about 3° further north (az. 117°). Photo: Lomsdalen.

was built. This would be comparable with observations made at the altar in Room 7 of the WSSR in 2010 and 2011 as noted earlier. Another potential astronomical alignment from the same area of this altar faces a major lunar standstill moonrise. This could be observed diagonally from the south edge of the altar along the south side of the entrance to Room 8 (Fig. 5.2). This orientation shows an azimuth of about 127° (declination -29.87°). This would make the most southerly moonrise during the Temple Period, which occurred at -29.2°of declination, visible.[5] I investigated this hypothesis and verified that the most southerly moonrise 4,000-5,000 years ago was indeed visible from the altar. Possibly the temple builders were symbolically marking the lunar cycle of 18.6 years, which could be related to the approximate period of a human generation.[6]

5.2 ARCHAEOASTRONOMICAL SURVEY OF MNAJDRA SOUTH TEMPLE

As this temple possesses the most complex structure of the Mnajdra compound the results will be presented separately for each part of the building. Some of the findings correspond to previous measurements as presented in Table 5.1; other retrieved orientations seem not to have been previously documented and will be discussed in the following sections.

---

5 Agius and Ventura, 'Investigation', p. 13.
6 Malville, *Prehistoric*, p. 38.

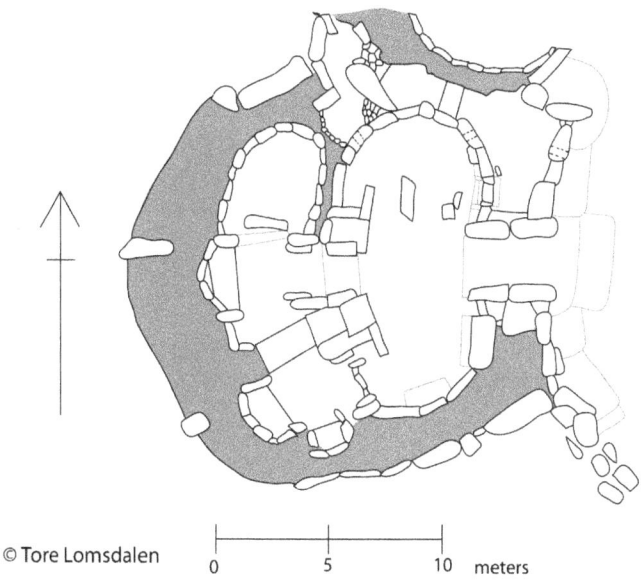

FIGURE 5.4. Plan of the South temple with its eastern orientation.[7]

5.2.1 MAIN ENTRANCE

The main entrance has the following coordinates: N 35° 49.600', E 14° 26.179'. A prominent feature of the Mnajdra site is its eastern horizon, located about 500 metres away. Towards the north, the hill has an altitude of 4.5° curving down to about 0° where the horizon meets the sea, as seen

---

7 Evans, *Antiquities*, Plan 20A.

from the South Temple; it continues towards the cliff that abruptly falls onto the shore. I have discovered no other megalithic temple on Malta or Gozo with such a horizon feature adequately situated for observing the path of the rising sun throughout the year (Fig. 5.5).

FIGURE 5.5. Horizon from the main entrance of the south temple with SSSR (left line), EQSR (central line) and WSSR (right line) points. The WSSR shows clearly that the sun rises at the apparent horizon point where land, sea, and the sky meet. Photo: Lomsdalen.

The main entrance of the South Temple permits observation of the sun's path along this horizon throughout the year. The open central orientation from the back niche through the temple entrance measured an azimuth of about 92° with a horizon altitude of about 4° (declination +0.7°). This closely matches the EQSR, and the illumination of the back altar at EQSR was photographically documented (Fig.

5.6). At WSSR the extreme part of the right-hand vertical orthostat was illuminated in a cross-jamb view and showed an azimuth of 120°, with a horizon altitude of 0° (declination -23.9°) (see Fig. 5.7). In symmetry with the WSSR, SSSR illuminates the outer part of the left-hand orthostat (also

in a cross-jamb view) with an azimuth of 63°, a horizon altitude of 4.5° (declination +23.6°) (see Fig. 5.8).

To investigate the chronology of the South Temple's building stages, a closer inspection of the main entrance was conducted in December 2012 to search for any signs of its having been extended. The shaft of the entrance is about 3 metres long, 1.8 metres wide and about 2 metres high. At about 1.9 metres from the outside are clear visible signs of a constructional split in both the threshold and side panels (Fig. 5.9). The outer vertical slabs are 1.9 metres long and 65–70 cm wide and consist of the harder Coralline Limestone; the inner ones are 1.1 metres long and 25–30 cm wide and the softer Globigerina Limestone has been used. This result will be discussed in section 6.2.4.

FIGURE 5.6. Above, EQSR illumination of the south temple's corridor; facing page, EQSR seen from the back altar. Photo: Lomsdalen.

5.2.2 HORIZON POSTHOLES

As mentioned in section 3.3, two postholes lying in the horizon of Mnajdra South were previously found by Ventura et al.[8] These cannot be dated, but their location in the landscape around Mnajdra suggests that they are linked with the temple complex.

The first posthole, identified with the help of Grima and Clive Cortis, was aligned with the WSSR as seen from Mnaj-

---

8 Ventura et al., 'Tally', pp. 172–73.

dra South and has coordinates of N 35° 49.748' and E 14° 26.119', about 353 metres distant from the temple entrance at 0° altitude. The hole is not completely circular and gives the impression of having been human-made, not machine-made. It has a diameter of about 35 cm and is equally deep. As measured from the entrance to Mnajdra South, a post inserted in this hole would have an azimuth of 120° (declination -23.92°). Although the sun rose about a sun disk (about ½°) further north of this position in 2010 (Fig. 5.7, right) it would have risen over the posthole during the

Temple Period when the sun had a declination of -24.0°.⁹ To examine this possibility, the experiment mentioned in chapter 4.4 was conducted on 26 June 2010 at 18:38 UT (the full moon rose four minutes later than listed in US Naval

FIGURE 5.7. Above, cross-jamb illumination at WSSR in the south temple. Facing page, view from the orthostat through the main entrance with the sun rising today about a sun disk further north than the pole on the WSSR horizon hole. Photos taken 17 December 2011 at 06:09 UT. Photo: Lomsdalen.

Observatory for that day at 18:34 UT). This moonrise was observed from inside the South Temple in a cross-jamb view, with an azimuth of 120° (declination -24.05°), only one twentieth of a degree south from where the sun would

---

9 Agius and Ventura, *Investigation*, p.13.

have risen five thousand years ago (Fig. 5.10). The observed alignment between the cross-jamb of the temple's entrances, the posthole, and the moon rising behind it was perfect.[10] I also noticed that the rising full moon casts no illumination inside the temple, as the amount of daylight is

---

10 Tore Lomsdalen, 'Possible Astronomical Intentionality in the Neolithic Mnajdra South Temple in Malta' (paper presented at the European Society for Astronomy in Culture [SEAC], Portugal, 2011).

still very strong just after sunset at 18:22 UT.[11]

The second identified posthole is at coordinates N 35° 49.691' and E 14° 25.455', about 345 metres distant from the South Temple entrance at a horizon altitude of about 4.5°. Thirty years after its discovery by Prof. Ventura's team, a small group and I searched for the posthole in the summer

---

11 USNO, 'Naval Oceanographic Portal', (http://www.usno.navy.mil/astronomy) [Accessed 20 June 2010].

of 2011. Authorised by the landowner and led by Ventura, the original hole was rediscovered. Like the WSSR posthole, this one also gives the impression of not being machine-made, is not completely circular in shape, and has a diameter of about 34 cm. It is 33 cm deep in the back and 25 cm deep in the front. Its general appearance and its distance

FIGURE 5.8. Facing page, SSSR's cross-jamb illumination. Above, SSSR sunrise as seen from the illumination point on the orthostat. Photos taken 23 June 2010 at 04:25 UT. Photo: Lomsdalen.

from the South Temple are remarkably similar to the first posthole. A post was also placed in this hole to allow it to be observed from Mnajdra. This confirmed that the posthole was about 3° further south than SSSR (Fig. 5.11).[12] This posthole's azimuth, when viewed from the entrance to Mnajdra South, is about 67° (declination +21.2°), whereas present-day SSSR occurs at about 64° of azimuth and a horizon altitude of about 4.5° (declination: +23.6°). During the Temple Period the SSSR had a declination of +24.0°, meaning that the sun rose about three quarters of a sun disk further north than it does today, even further away from the posthole.

Attempts were made to find a possible hole more precisely aligned with the SSSR. This should, theoretically, be about 25 metres further north, as confirmed on site with a handheld GPS. Due to heavy vegetation it was not possible to conduct a thorough search and the land would need to be cleared to find such a hole, if it exists.

5.2.3 CROSS-QUARTER AND EIGHTH DAYS

To test Foster's and Micallef's assertions that the full arc of the sun throughout the year was observed from Mnajdra South, I made observations of the cross-quarter and eighth days throughout a 12-month period.[13]

---

12 Lomsdalen, 'Intentionality'.
13 Thomson Foster, 'Orientation'; Micallef, 'Alignments'.

FIGURE 5.9. Main entrance to the south temple: on the left is the south side and on the right, the north side. Both vertical orthostats indicate variations in construction techniques and the threshold seems to have added slabs in the same areas as a possible extension of the corridor (see chapter 6.2.4). Photo: Lomsdalen.

FIGURE 5.10. Moonrise on 26 June 2010 at 18:38 UT over the WSSR posthole at declination -24.05°. Photo: Lomsdalen.

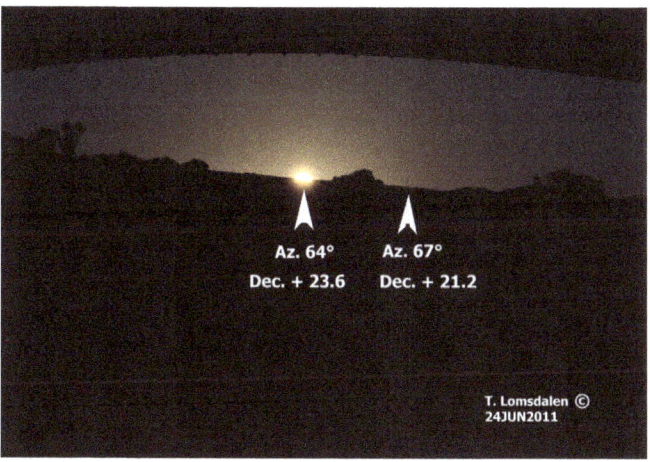

FIGURE 5.11. SSSR indicating actual sunrise about 3° further north than the marker on the assumed horizon portal. View from the main entrance of the south temple on 24 June 2011 at 04:22 UT. Photo: Lomsdalen.

As previously discussed and illustrated, EQSR illuminates the central corridor and the back altar, while the WSSR and SSSR illuminate the outer edges of the vertical orthostats in a cross-jamb view. The cross-quarter-day sunrise illuminates the inner edges of the northern orthostat in February and November, and the southern orthostat in May and August. The eighth-day sunrises, which occur halfway between the equinox and cross-quarter days, produce offset illuminations that are equally symmetrical. The sunrises in February and October illuminate the beginning of the northern horizontal altar slab; sunrises in April and August illuminate the beginning of the corresponding southern altar. The eighth days midway between the solstices and the cross-quarter days display the same form of symmetry as the others. In January and November the sun's first rays illuminate the northern orthostat about one-third from its outer edge; in May and July a similar image is impressed on the southern orthostat (Fig. 5.12). Travel issues prevented

FIGURE 5.12. Constructed image of slit illuminations of sunrise throughout the year. Photo: Lomsdalen.

observations of all sunrises at the cross-quarter and eighth days, but when I was not present, photographic documentation was taken by an assistant, although full documentation of the azimuth, horizon altitude, and declination of sunrise was not measured for these events.

The following table indicates when the various Cross-Quarter and Eighth Days, Solstices, and Equinoxes took place for the field study in question. All months are abbreviated to three letters.

DATE	SOLAR EVENT
22 DEC 2011	WINTER SOLTICE
13 JAN 2012	EIGHTH DAY
05 FEB 2012	CROSS-QUARTER DAY
28 FEB 2012	EIGHTH DAY
20 MAR 2012	SPRING EQUINOX
13 APR 2012	EIGHTH DAY
06 MAY 2012	CROSS-QUARTER DAY
30 MAY 2012	EIGHTH DAY
21 JUN 2012	SUMMER SOLTICE
14 JUL 2012	EIGHTH DAY
06 AUG 2012	CROSS-QUARTER DAY
31 AUG 2012	EIGHTH DAY
23 SEP 2012	AUTUMN EQUINOX
15 OCT 2012	EIGHTH DAY
06 NOV 2012	CROSS-QUARTER DAY
28 NOV 2012	EIGHTH DAY
21 DEC 2012	WINTER SOLSTICE

TABLE 5.2. Cross-Quarter and Eighth Days in 2012.

5.2.4 ORACLE HOLES

In the temple wall of the north apse are two holes that have tentatively been interpreted as oracle holes (see chapter 2.4.2 and Fig. 2.14.). The larger hole on the right, where a portal doorway with evident rope holes for a door leads into

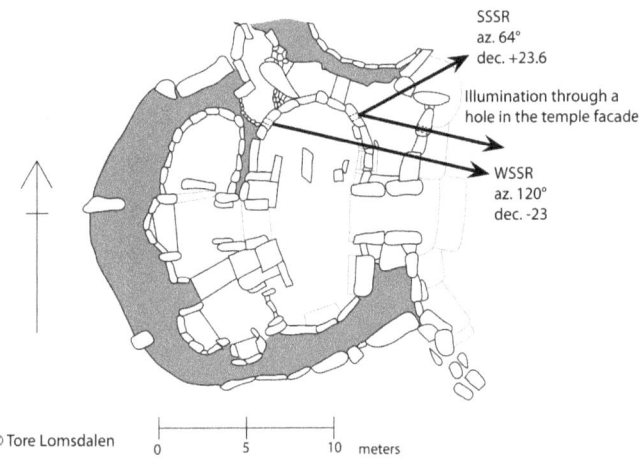

FIGURE 5.13. Alignments from oracles holes towards WSSR and SSSR in Room 5 and 6.

Room 5, was measured in a cross-jamb view through the hole from the apse side; the measured azimuth of about 64° corresponds with an alignment towards the SSSR (Fig. 5.13). This hole also receives offset illumination from the WSSR through a hole in the front façade, as first noted by Agius and Ventura.[14] As described in section 4.1.2 this event was photographically documented at WSSR 2012 currently with the WSSR seen from Room 1 as the sun rises over the horizon posthole (Fig. 5.14). On the side of the oracle hole between Room 5 and 6, a now-collapsed altar-like niche would also have received illumination as the sun rose higher in the sky.[15] The smaller oracle hole on the left side of the apse with only an entrance from the backside of the temple (Room 6) was measured from inside the room and out into the apse. It shows a cross-jamb azimuth of about 120°; it is thus oriented towards the WSSR, an observation that had not previously been mentioned (Fig. 5.13).

5.2.5 OTHER POSSIBLE ALIGNMENTS

In addition to the three main alignments visible from Room 1 of this temple (Fig. 3.9), I measured orientations from other areas of the temple. Room 2 is placed along the main axis of the South Temple and is illuminated by the sun's rays at EQSR, as previously stated. Additional measurements were made from both the north and south sides of the altar in a cross-jamb view through the entrance to Room 2; orientations to both SSSR and WSSR with azimuths similar to the main entrance features were found. These had not been considered by previous scholars (Fig. 5.15).

---

14 Agius and Ventura, 'Investigation', fn 14, p. 32.
15 Thomson Foster, 'Orientation', Figs. 15–17.

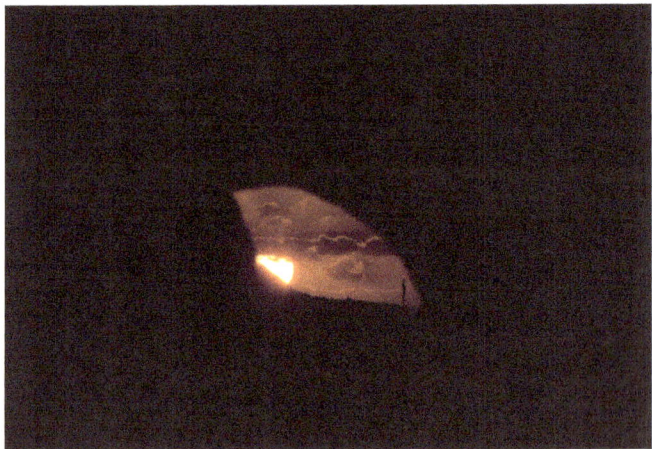

FIGURE 5.14. Above, WSSR illumination of the oracle hole in Room 5; on the right in the photo, the collapsed altar. Below, a photo taken sitting in front of the illumination point, observing sunrise over the WSSR posthole. Photo: Lomsdalen.

From the left and right altars on each side of the central corridor in Room 1, I discovered two new alignments to WSSR and SSSR (Fig. 5.16). These cross-jamb illuminations would be similar to the WSSR in the middle temple (see chapter 3.3, Figs. 3.6 and 3.7) if the outer part of the south temple's façade was not yet constructed, a hypothesis discussed in chapter 6.2.4.

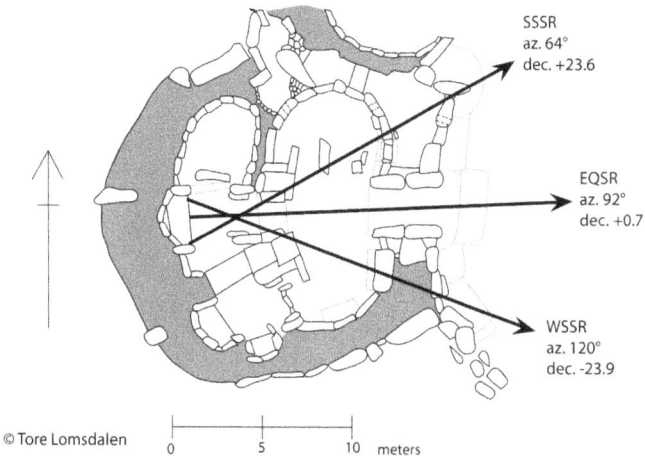

FIGURE 5.15. Alignments from back altar Room 2.

In Room 3, at the back left-hand niche (Fig. 5.17) there are three orientations towards the equinoxes and solstices. From the centre of the middle niche through the centre of the portal entrance was measured an azimuth of about 63°, corresponding to the SSSR. From the north side of the same niche, through the centre of the doorway was measured an azimuth of about 92°, roughly equal to EQSR. From the centre of the north niche in a cross-jamb view through the portal entrance was measured an azimuth of about 120°, representing the WSSR. At the time of research into these previously unpublished alignments, I was unaware that in 1999 Thomson Foster had indicated a SSSR alignment from

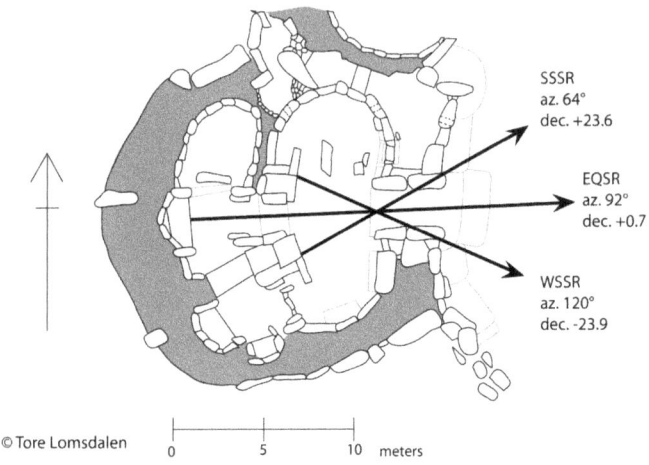

FIGURE 5.16. Alignments from the left and right altars in Room 1.

the very spot I suggested it to be inside Room 3; however, she did not indicate any WSSR or EQSR alignments from this room.[16] In addition it is not clear whether Thomson Foster's Figure 18 indicates a theoretical prolongation of the sunlight at SSSR from Room 1 to 3, or a separate point of observation of the sunrise. Based on today's architecture,

---

16 Thomson Foster, 'Orientation', Fig. 18.

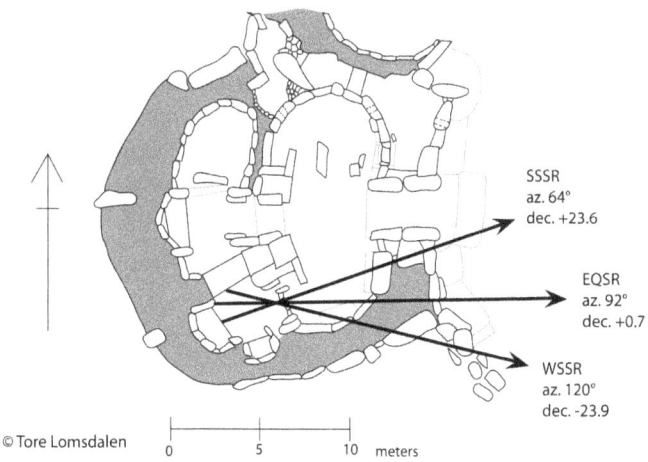

FIGURE 5.17. Alignments from Room 3.

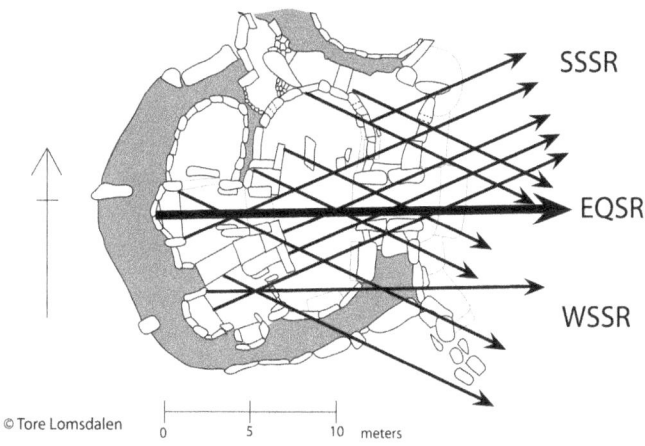

FIGURE 5.18. Mnajdra South and the sunrise alignments I investigated.

Room 3 cannot receive any direct illumination at SSSR, WSSR, or EQSR.

My investigation of the South Temple suggests, in sum, twelve alignments towards the sunrise at the equinoxes and solstices; five or possibly six seem not to have been previously considered or documented (Fig. 5.18). Regarding the North Temple, three alignments toward the rising sun and the moon were found, two of which had not previously been documented (Fig. 5.2).

This chapter has presented my fieldwork results, beginning with a comparative listing of the orientations of the following temples on the Maltese archipelago: Ġgantija South, Ġgantija North, Ħaġar Qim North, Ħaġar Qim Main, Mnajdra South, Mnajdra Middle, Mnajdra East, Skorba E, Skorba W, Ta' Ħaġrat I, Tarxien East, and Tarxien Main. However, the main emphasis of the fieldwork focused on documenting cardinal, cross-quarter and eighth day solar illumination patterns found in the Mnajdra Temple compound. The following chapter will discuss the implications of these results.

CHAPTER 6

DISCUSSION

IN THIS CHAPTER the fieldwork results from Chapter 5 are discussed and integrated into the wider debates surrounding prehistoric Malta and, specifically, the Mnajdra temple complex.

6.1 MALTESE ARCHAEOASTRONOMY

As Table 5.1 shows, my surveyed measurements are consistent with those of previous authors, indicating that, firstly, the employed fieldwork methodology was sufficiently accurate, especially considering that Agius and Ventura used a theodolite in their 1981 surveys of the temples' axes. This coincides with studies by Ventura *et al.*, which show consistent southeast to southwest temple orientations, which 'cannot have come about by chance'.[1] Agius and Ventura also conclude that 'some factor has influenced the choice of orientation of the temples' axes'.[2] Cox concludes that temples of the Ġgantija Phase are aligned along a NW-SE axis, while temples of the Tarxien Phase are aligned along a NNE-SSW axis.[3]

From my own observations of the orientations of the five best-preserved temple complexes on the archipelago, four temples (Ġgantija, Ta' Ħaġrat, Skorba, Ħaġar Qim, and Mnajdra North) seem to receive a cross-jamb illumination through their central entrances onto a left-hand side altar arrangement at winter solstice sunrise. The only outlier is Mnajdra South, which receives this offset illumination on the right-hand side. The Tarxien temple complex is also well preserved but its complexity requires additional research. However, Tarxien temple's main direction is southwest, possibly aligned to winter solstice sunset.

Based on my site visits and measurements I argue that, in general, the purpose of the central axis' orientation was to create offset illumination effects that directed sunlight onto demarcated areas within the temple at specific times of the year. By this I mean that merely illuminating the temples' centre or central corridor was not the builders' main priority. I argue that they based their temple architecture on more sophisticated alignments that screened

---

1 Foderà Serio *et al.*, 'Orientations', pp. 115–17.
2 Agius and Ventura, *Investigation*, p. 9.
3 Cox, 'Orientations', p. 36.

solar rays in order to illuminate key areas, leaving other parts in darkness. My argument builds on Vassallo's claim that sunlight must have been a very important factor for the early temple builders.[4]

Although my sample size is limited, these results also suggest that the yearly WSSR prevailed in importance over other sunrises throughout the year. This idea is based on Vassallo's broader study, which concludes that 'a total of twenty-one of twenty-four sites (or 88%) shows an alignment with the winter sunrise'.[5] In modern times, we are clearly aware that at winter solstice, when the sun reaches its most southern position in the sky of the northern hemisphere, there is a change as the days become longer, the temperature gradually rises and most of us look forward to more agreeable weather. One cannot exclude the idea that prehistoric peoples were also aware of this seasonal shift, and welcomed it by creating such special atmospheric effects as offset illumination, feasts, rituals, and offerings.

6.2 INTENTIONALITY BEHIND MNAJDRA

The complexity behind this temple structure requires a separate analysis and discussion of each element, before attempting any holistic conclusions. For this I will discuss the observed offset illumination effects, the relationship between the oracle holes and astronomy, the linkage be-

---

4 Vassallo, 'Sun Worship'.
5 Vassallo, 'Location', pp. 44–46.

tween the postholes and the temples and, finally, the role astronomical alignments seem to have played in the various phases of construction of the temple complex.

6.2.1 OFFSET ILLUMINATION AND THE LIGHT/DARK DICHOTOMY

Based on the many site visits to especially Mnajdra South and Mnajdra Middle, the solar illumination at certain times of year does seem to be intentional. Based on today's architecture (for chronology, see section 6.2.4), any site visitor can observe sunrise illumination through the main entrances onto demarcated interior areas at the solstices and equinoxes. Particularly at Mnajdra South, the crossjamb illuminations at WSSR and SSSR are not only an equal distance from the temple's central axis, but also throw their light on the very outer edge of pit-decorated orthostats. This does not occur in an absolutely identical manner, but it is nevertheless close enough for a viewer to experience symmetry. The cross-jamb view is so narrow that illumination lasts only a few minutes. Whether such a feature was purposely constructed has so far not been proved, but its exactness may indicate the intentionality of its nature.

Such a feature further indicates that the temple architects purposely screened the areas they wanted illuminated and the ones to be kept dark, which was suggested above as a general rule for the Maltese temples. Another probable example of intentional light and dark effects can be seen at the EQSR, when the central corridor of the temple is fully flooded by the sun's rays, while the side apses remain in complete darkness. In addition, at sunrise the sun's rays never reach above the horizontal slab of the back altar at the end of the corridor in Room 2. On spring equinox in 2011, I placed an object on top of the altar which never received any light as the illumination 'shaft' moved slowly from left to right along the central corridor. Thus, any objects kept on this altar would likely never be illuminated by the sun throughout the year (Fig. 6.1).

FIGURE 6.1. An object on the top altar plate does not receive illumination. Photo: Lomsdalen.

Mnajdra Middle also has a demarcated cross-jamb illumination through the central portal entrance, as the left-hand altar of the central corridor receives light at sunrise on the winter solstice. On the left side of the temple is an engraved figure (Fig. 2.9, left) which looks like a roofed temple, of unknown date, that receives the first rays of the morning sun. As Mnajdra North stands today, the illumination shaft moves from the left-hand side altar to the right-hand side, mirroring the movement of the rising sun that eventually disappears behind the protective covering tent (erected in 2009). Without this tent the whole of the temple would, at certain times of day, be bathed in sunlight; consequently, demarcated light/dark areas would not be possible except just at sunrise—unless the temple was roofed (Fig. 6.2).

FIGURE 6.2. Middle temple illuminated at WSSR on 25 December 2012 at 06:26 UT. Photo: Lomsdalen.

FIGURE 6.3. Artistic impression of a hypothetical roof on the Mnajdra South Temple. Drawing by Anna Grima.

In 1939 Ceschi made an artistic impression of how Mnajdra South may have looked if roofed.[6] The Maltese artist Anna Grima has given another version of how the south temple roofing may have been (Fig. 6.3).[7] Both Evans and Trump

---

6 Carlo Ceschi, *Architettura dei Templi Megalitici di Malta* (Rome: Casa Editrice Fratelli Palombi, 1939), Fig. 34.

7 Anna Grima, 'Maltese Temple Roofing' (Private collection, 2014).

take into consideration that the Maltese temples may have been roofed.⁸ This idea is supported by the representation of a temple façade, believed to be a 'temple building model' from the Temple Period, retrieved from Ta' Ħaġrat, which shows roofing (Fig. 2.9, right).⁹ Whether the temples were roofed is still an open question. But it is important to note that, if the temples were roofed, the sensory experience of light and dark would have been a major feature of these temples.

The light and dark dichotomy at sunrise may have been intentionally created for religious and sacred performances. At the moment of sunrise illumination is limited to discrete areas, which could indicate that the temple's inner space was meant to accommodate only a few people. Since Renfrew proposes that Maltese pre-historical society was governed by a hierarchical chiefdom, it is possible that a priestly class conducted ceremonies and rituals for the select few inside, while the masses observed only part of the ceremony from the forecourt outside.¹⁰

6.2.2 AN ASTRONOMICAL INTENTION BEHIND THE ORACLE HOLES IN MNAJDRA SOUTH

As far as I have been able to ascertain, the oracle holes of Mnajdra South do not seem to have been directly exposed

---

8 Evans, *Malta*, p. 126; Trump, *Malta: Prehistory*, p. 150.

9 Pace, 'Sites'.

10 Renfrew, *Civilization*, p. 170.

to the rising sun's rays, except the right-hand hole, which receives WSSR light through a hole in the façade (Fig. 5.14). If the temple builders had a celestial purpose or special intention for the oracle holes, it could have been symbolic rather than observational. As one oracle hole is directed both towards SSSR and WSSR, and the other towards WSSR, it can be argued that their placement might be intentional. Each hole might have been used for rituals and ceremonial activities at times corresponding to events occurring at the sun's most northern and southern positions on the horizon, at the spring and autumn equinox, or at other special times related to astronomical events and their cosmology. Such a feature implies the temple builders' ability to work with symbolism in addition to the physical phenomenon of the sun's light.

Although uncertain, if the word 'oracle hole' carries any factual significance or symbolic value as suggested by other authors, it can be argued that the holes were used for passing valued objects, meaningful sounds and advice, either in combination with or separate from, religious practices and rituals.

6.2.3 THE SOLSTITIAL POSTHOLES

The solstitial postholes are two essential factors in judging whether Mnajdra South was intentionally aligned to the rising sun at the solstices. Results from astronomical investigations done by Ventura *et al.* and myself favour the idea that the winter solstice posthole may have been intentionally created for alignment purposes; it is precisely aligned to the spot where the sun would have risen about 5,000 years ago.[11] However, Ventura *et al.* maintain that the theory of astronomical intentionality is marred by the fact that the sssr posthole is 3° off the sun's rising point.[12] Furthermore, the wssr point focuses on another distinctive aspect in the landscape, as seen from Mnajdra: it marks the location where the land meets the sea and the heavens. This phenomenological observation cannot be simply dismissed; all three may have been important components of the prehistoric islanders' cosmology, as remarked by Grima (chapter 3.1). I, myself, noted and experienced that observing a celestial body rising over this position, which at the same time creates an illumination effect inside the temple, produced a strong sensory and emotional effect. One can only imagine what it might have been like for a person from the Temple Period whose cosmological awareness was more closely attuned to the topography of the area.

The other posthole, as noted, is too far south to be used as a marker for the sssr as seen from Mnajdra South. However, one cannot exclude the possibility that it might have been used for other purposes, or that another posthole, more correctly aligned to the sssr, is still to be found. The question of whether the postholes were dug out by the temple builders, or whether they are younger and intended for purposes unrelated to the temple, is therefore, still open.

---

11 Ventura *et al.*, 'Tally'.
12 Ventura *et al.*, 'Tally', p. 178.

6.2.4 THE CONSTRUCTION SEQUENCE OF MNAJDRA

Chapter 5 mentioned other possible alignments to the equinoxes and solstices found in rooms of Mnajdra South that do not receive illumination. These could be accidental, or merely symbolic, but my hypothesis is that they once provided offset illumination just as the main entrance now does. As noted in chapter 2, the Mnajdra complex was not built as a whole: rooms and features were added on in several stages. The observed orientations could, then, have been 'live' at some stage. In fact, archaeoastronomy may assist archaeology in the identification of the construction sequence.

Astronomically, Room 3 of Mnajdra South has alignments to the equinoctial and solstitial sunrises, during which specific areas inside would be illuminated (Fig. 6.4, left). These alignments are not as precise and demarcated as later parts of the South Temple and may indicate an earlier period of horizon-based astronomical knowledge. The extension of this temple might have first been towards Rooms 2 and 4. Astronomically, Room 4 seems to be of little interest; however, Room 2 is highly central to the temple's overall axis of about 92° azimuth (declination +0.7°) which aligns with the EQSR, fully illuminating the altar at the back niche. The Room 2 altar would also receive cross-jamb illumination on solstitial sunrises through its entrance, if Room 1 was not yet constructed (Fig. 6.4, centre).

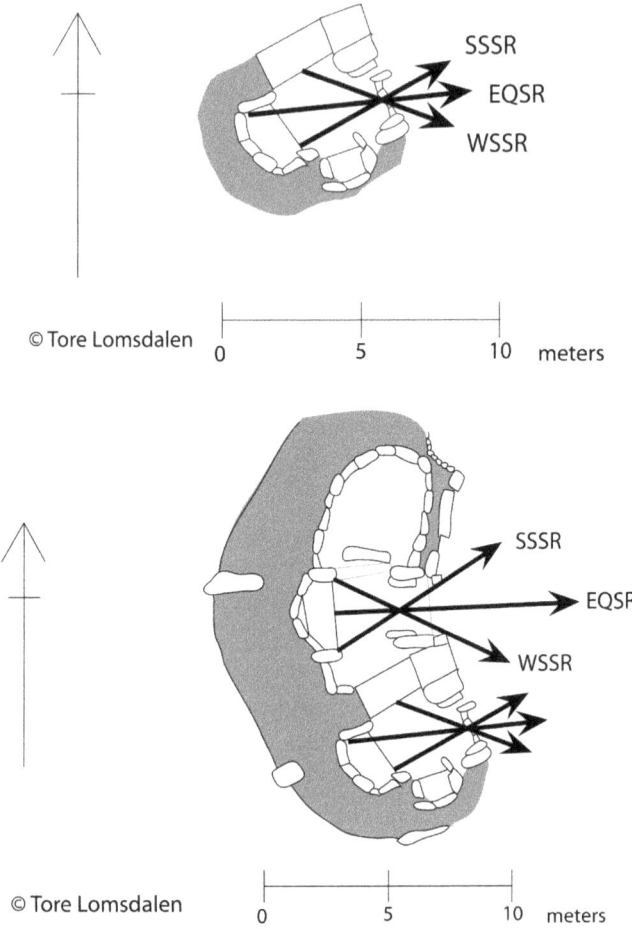

FIGURE 6.4. The suggested first three building stages of the south temple. Above, Room 3; below, adding on Rooms 2 and 4; facing page, the extension of Room 1. All have alignments towards EQSR, SSSR and WSSR.

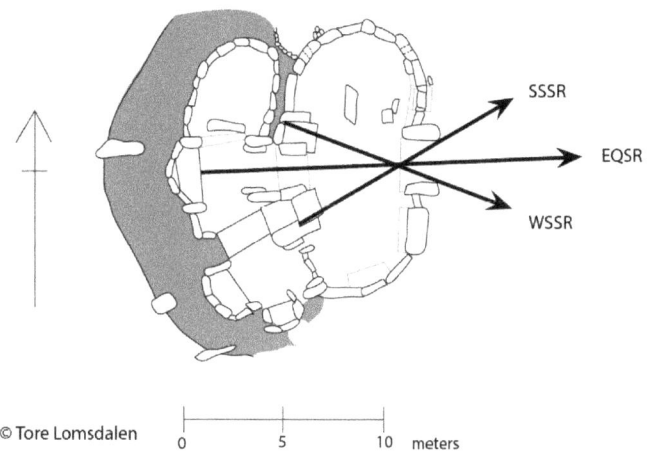

© Tore Lomsdalen

The next construction phase would be the extension to Room 1 and the front apses, which seemed to be a normal temple building procedure (as previously mentioned). Room 1 is similar to the later-built Room 7, in which there is an altar on each side of the passage into the back room. At sssr and wssr, the two altars in Room 1 would have received a cross-jamb illumination before the concave façade and the extension of the main entrance was built (Fig. 6.4, right). A close inspection of this entrance suggests that it was built in two stages (see chapter 5.2.1): at about ⅔ along its length is a split in the threshold and the side walls, where different sizes and quality of panel slabs were used, giving the impression of a second-phase extension of the entrance towards the forecourt. I have been unable to find any archaeological documentation to substantiate or reject this observation. The rear part (Room 8) of the North

Temple might have been constructed during the same time period or shortly afterward (Fig. 6.5, left).

With the building of Room 5, the completion of the North Temple (Room 7), the creation of the concave façade of the South Temple, and the elongation and narrowing of its main entrance, the two altars just mentioned would be closed off from the solstice sunlight (Fig. 6.5, right). I suggest that this was when the two vertical orthostats were placed on each side of the altars in order to receive the sol-

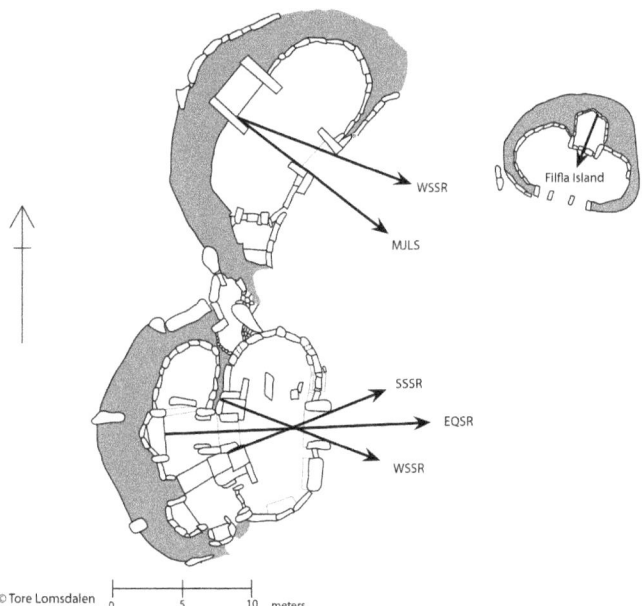

FIGURE 6.5. Above, the suggested fourth building stage adding Room 8 of the middle to the south temple. Facing page, the completion of the Mnajdra complex with alignments.

Discussion 149

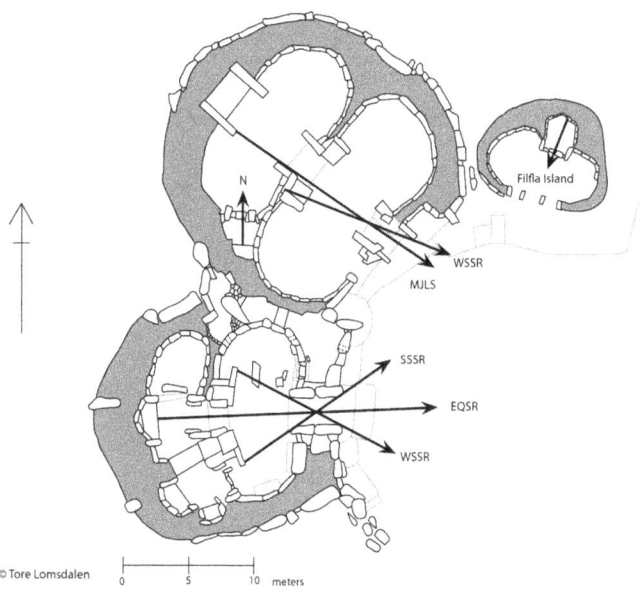

stitial cross-jamb illumination, as can be observed today.

Pace states that Room 5 was 'fashioned out of the wall of the Lower Mnajdra' with the former external megaliths of the older building being used as the supporting wall of the middle temple.[13] This would have been done in the Tarxien Phase. It seems that Pace assumes that Mnajdra South was fully completed before the builders started to erect the middle temple. However, I argue that Room 5 came into existence as a consequence of the extension of Mnajdra Middle from a two- to a four-apse temple when Room 7 was

---

13 Pace, 'Sites', p. 131.

built. My argument agrees with Evans' findings of pre-Tarxien types of pottery under the floor of Room 5, suggesting that there was a building on this site in the previous Ġgantija Phase.[14] Evans' indications may suggest that this

FIGURE 6.6. Room 5 seen from the north-east corner indicating an outer dressed wall of Room 1 and a clear distance between the south and middle temple with the remains of the collapsed altar in the middle. Photo: Lomsdalen.

---

14 Evans, *Antiquities*, p. 103.

area was not filled with wall packing before Room 5 was built and consequently could have been created through extension of the main entrance and the concave façade. In addition, my observations suggest that the wall separating Room 5 from Room 1 appears to be a dressed outer wall of Room 1. If the builders intended the apse wall as a part of an area filled with earth, it may be assumed that less labour would have been put into aesthetic work of that side of the wall; neither wall gives the south temple any impression of being a supporting foundation for the middle temple (Fig. 6.6).

The question of the three Mnajdra buildings' chronology and a verifiable origin of constructional authenticity is challenging, as suggested in chapter 2.4.3. However, my argument for a redefined building sequence is based on archaeological and archaeoastronomical components. As described in chapter 2.4.2, the small east temple seems to have undergone heavy reconstruction work in modern times, but this falls outside the scope of astronomical alignments investigated for this book.

Regarding the middle temple, Evans states 'one cannot doubt that it was all constructed at once and not subsequently altered' due to its homogenous construction throughout (chapter 2.4.2).[15] But for Mnajdra South, chapter 2.4.2 describes both the archaeological examinations, which suggest various building stages, and scholarly differences in opinion regarding the building sequence of the various chambers and apses. Old photographs do suggest

---

15 Evans, *Antiquities*, p. 102.

a certain disorder of the temple as mentioned in chapter 2.4.2; nevertheless, Evans points out that the façade has an archaic look and the corridor is well preserved.[16] Therefore, I argue that the fundamental base and core construction seem to be original, rather than the result of later reconstruction efforts.

In accord with the archaeological finds from each room (detailed in chapter 2) and my own archaeoastronomical investigations, I propose the following refinements to the constructional sequence:

1. Mnajdra East might have been the first to be constructed, in the early Ġgantija Phase (3,600-3,000 BCE). This idea is rooted in typological and archaeological considerations; unfortunately, archaeoastronomy can add little.

2. Mnajdra North could have been built in two separate stages, one in the middle and the other in the late Tarxien Phase (3,000-2,500 BCE). Based on archaeoastronomical observations, its construction could have started with the back apses (Room 8) and later been expanded by adding a new apse (Room 7) to the temple.

---

16 Evans, *Antiquities*, p. 96; Pace, 'Sites', p. 129.

3. The complex Mnajdra South may have been constructed in four stages:

   1. Room 3 dates from the early Ġgantija Phase and could be contemporary with, or even older than, Mnajdra East. It possesses several characteristics of a temple in its own right, including the solstitial and equinoctial illumination displayed by the final version of the South Temple.

   2. Rooms 2 and 4 could have been added in the middle Ġgantija Phase. Room 2 again replicates the same archaeoastronomical signature now in its final form (with the central axis oriented towards the East).

   3. Extending the temple with the front apses (Room 1) seems to have been the third building stage and may have been completed sometime in the late Ġgantija or early Tarxien Phases.

   4. The fourth and final stage may have been the erection of Room 5 as a foundation support for the Room 7 extension of Mnajdra North. To maintain architectural uniformity the entrance to Mnajdra South was then elongated and its present concave façade established. With this, fifteen hundred years of off-and-on building was concluded, already well into the Tarxien Phase.

In this chapter I have examined and discussed the results from my investigations of a possible astronomical intentionality behind the architecture of the Mnajdra Temple complex from its first stages of construction to its completion one and a half millennia later. In the following chapter I will draw together all the facets of this research project.

CHAPTER 7

CONCLUSION

THIS CHAPTER SUMMARISES the arguments of my research in Malta and the major findings regarding the hypothesis of a sky-based intentionality behind the construction of the prehistoric Mnajdra Temple complex.

The prehistoric temple builders' astronomical purposes or intentionality cannot be verified as there is no written documentation to support such an assumption. Therefore, all evidence is circumstantial, but should not be dismissed merely because it is difficult to quantify. As mentioned in this book there are, both inside and outside Malta, substantial indications that prehistoric societies' awareness of astronomical phenomena influenced human behaviour.

My objective was to search for evidence of intentional sky involvement when constructing the temples at the

Mnajdra site, inspired by Trump's reluctant argument of an astronomical intentionality behind the architecture of the temples (chapter 3.2). The investigation focused on whether there appeared to be a continuous interest in relating the sun's position to the temple structures throughout their various building sequences. Such evidence would have shown consistency in considering the sun's position on the horizon throughout the temples' one and a half millennia of construction and usage. My results offer new insights into questions that surround these temples. The search for celestial intentionality behind their architecture also suggests a proposed redefined building sequence. The following is a consideration of each temple individually.

The Mnajdra East temple, due to its central southwest orientation towards the Filfla islet, is the only temple on the site without an orientation towards sunrise. The temple may owe a stronger allegiance to Filfla than to the sun. Filfla may have been a site of both religious and cosmological significance to the temple builders, as Temple Period remains have been found there.[1] The islet's altar-like profile on the skyline may have been an attractive feature, as altars seem to have been an important religious element in temple constructions. Besides that, as seen from Mnajdra East, the islet is on a meeting point of earth, sea, and heaven—all three being important ingredients of an island people's cosmology. One of the apses at Ħaġar Qim, Mnajdra's next door neighbour, is also oriented towards Filfla, which strengthens claims of an intentional alignment with the islet from

---

[1] Farrugia Randon, *Filfla*, p. 43.

both temple sites, as well as the significance the builders may have attributed to it. Ħaġar Qim and Mnajdra are the only two temples in the islet's vicinity. I therefore argue that Filfla had cosmological significance for the temple builders.

From Ventura *et al.*'s suggestion that the tally stones of Mnajdra East are a genuine counting device it would appear that the temple builders were keen observers of celestial bodies.[2] If so, claims of intentional astronomical orientations of the other temples are strengthened. It further vitalises the theory that Malta's prehistoric society may have employed a calendric division of time, possibly for agricultural purposes or to establish occasions for religious feasts, rituals and ceremonies. That the tally stones may have been an arithmetic tool for finding the equinox by counting the days between the two solstices is not provable; nevertheless, the hypothesis should not be disregarded, especially since the tally on the east pillar has 179 holes, close to half the number of days in a year (from a winter solstice to a summer solstice).

Mnajdra Middle is a more astronomically complex temple than its smaller eastern neighbour. To address the question of astronomical intentionality regarding this temple two factors have been considered: the offset illumination produced at winter solstice sunrise and the orientation towards the major lunar standstill moonrise (Fig. 5.2). The most significant feature seems to have been the alignment towards winter solstice sunrise—as suggested by Vassallo—

---

2 Ventura *et al.*, 'Tally'.

which prevails in 88% of temple orientations throughout the archipelago.[3] However, in order for a cross-jamb illumination to take place, one needs both the correct solar alignment and internal temple geometry from the portal entrance to the side altar to be illuminated at sunrise. The axial orientation towards the major standstill moonrise may be epiphenomenal, according to my own observation; the full moonrise casts no light into the temples. Alignments towards the most southerly moonrise are also found at Ħaġar Qim, Ta' Ħaġrat, and Ġgantija.[4] However, it is a more likely assumption that the sun was the builders' main focus.[5]

Mnajdra South is the temple that has generated much enthusiasm, research and opinions, as its present architecture allows symmetrical interior illuminations at both equinoctial and solstitial sunrise. It thus gives the impression of being purposely aligned and oriented towards the passage of the rising sun along its eastern horizon. To determine whether this was intentional, four key aspects need to be considered: the offset illumination at sunrise, the existence of postholes at key locations on the horizon, Mnajdra South's unique architecture among the Maltese temples, and the building sequence of this complex site.

---

3 Vassallo, 'Location'.
4 Cox, 'Observations of Far-Southerly Moonrise'.
5 Vassallo, 'Layout'.

OFFSET ILLUMINATION

The equinoctial orientation and cross-jamb illuminations at the solstices are the most recognizable features of Mnajdra South. Some authors are reluctant to acknowledge astronomical intentionality behind these features, instead suggesting that they are due to chance. However, the internal façade of this temple also marks the other vital solar days of the year, the cross-quarter and the eighth days. I found that, at dawn on these days, as well as on the solstices, key thresholds of the side altars and the vertical slabs are illuminated by the offset rays of the rising sun, which suggests that the dimensions of the elements of this internal façade were carefully chosen to mark solar alignments. Thus I argue that there are too many alignments for them to be considered coincidental. My research leads me to the view that the temple's eastern orientation and architectural layout, as it presents today, seem to have been a consequence of astronomically based intentionality on the part of the builders.

POSTHOLES

Although the origin and dating of the postholes is unknown, the first posthole to be found is perfectly aligned with the winter solstice sunrise as it would have been seen from Mnajdra South during the Temple Period, confirmed by my observation of a moonrise with the same declination (Fig. 5.10). This supports claims of intentionality as the posthole could have been used to mark this horizon position during the temple's construction and/or for religious purposes (possibly by erecting a pole with a flag or totem). Admittedly, this claim is weakened by the imperfect alignment of the other posthole with the summer solstice sunrise. Another posthole, more perfectly aligned, might still be found; at present, the question of why only the winter solstice would have been marked by a posthole remains open. Nevertheless, even with difficulties of dating, the winter solstice posthole does strongly suggest intentionality.

UNIQUE ARCHITECTURE

Mnajdra South is the only extant Maltese temple with a clearly defined eastern orientation for its central axis. Other than that, it has all the essential characteristics of the other temples on the archipelago, except for its sophisticated offset solar illumination. This could have been the result of an increasing awareness by its builders of horizon-based astronomy, interest in finer fractional divisions of a solar year, and/or improving technical skill. This must all be kept in mind as one looks at the various stages of construction of this temple.

BUILDING SEQUENCE

Reports by archaeologists from the late nineteenth century until today exhibit a certain consistency in the theorised building sequence of the three distinct temples on the Mnajdra site. However, the more detailed analyses of the various rooms and apses there seem to produce more uncertainty and variations in scholarly theories of chronology. In archaeoastronomical research conducted on the site, previous scholars seem to have based their observations and interpretations on the assumption that the temples were originally built as they stand today. However, the archaeological finds, and my own astronomical measurements on alignments and field observations, led me to a different hypothesis, described in section 5.2.5. Astronomical considerations and intentional alignments to celestial bodies may have been major motivating factors in the construction of the very first part of the temple complex, as well as in its further extensions. The perceivable dichotomy of light and dark, created by the offset sunrise illumination, seems to have been an important effect even at the very early construction stages of Mnajdra South. This observation strengthens the religious, sacred and ritual implications of an astronomical and cosmological context where the sun's rays had an animating force of vitalizing life through the sun's cyclical oscillation from one extreme to the other along the horizon.

The argument for intentionality behind the astronomical alignments of Mnajdra is strengthened by the fact that these alignments were present throughout the various stages of its construction; thus, claims that they are solely due to chance lose substance. Furthermore, if the construction sequence proposed in chapter 6 is valid, then not only was the 'archaeoastronomical signature' of Mnajdra maintained throughout its several stages, it was improved upon and expanded. The technical complexity of the alignments of Mnajdra South, with its equinoctial orientation and solstitial, cross-quarter and eighth days offset illuminations, may be seen as the end result of a sequence of improvements over one and a half millennia, not as a one-off phenomenon, appearing *ex nihilo*.

This research began as a search for intentionality of solar alignments in the construction of the Mnajdra Temple on the island of Malta. Evidence for this intentionality also suggests new evidence that adds to the understanding of the actual sequence of the construction of the temple. More research needs to be conducted, particularly in seeking archaeological evidence on the chronology of temple construction, as well as the elusive posthole for the summer solstice sunrise position, but this will not refute the archaeoastronomical intentionality of the temple builders.

Appendix I:

A talk with Frank Ventura on astronomical observations related to Maltese prehistoric temples.

DATE: 22/03/2011

TORE LOMSDALEN (TL): I'm at the Malta University with Professor Ventura. Twenty second of March 2011. I understand, Professor Ventura, you are a Mathematician, right?

FRANK VENTURA (FV): I am actually a Chemist, I teach Science Education, so I prepare Science teachers to teach science and that means Physicists, Chemists and Biologists.

TL: How did this interest in astronomy come in?

FV: Oh, a long time ago. I was interested in astronomy since I was...the first I remember...I was eleven years old, so a young boy, and I got interested in astronomy. But then I got interested into Archaeoastronomy. When I read an article by Alexander Thom. I don't know if you are familiar...

TL: Yes, I know.

FV: And he had written after the publication by Gerald

Hawkins, *Stonehenge Decoded*.

TL: 1960.

FV: Exactly, this around about 1963. In 1963, I saw this article. He put in some diagrams of stones and he spoke about it…But it was in the 1960s that I had the first idea that something needs to be done about the temples.

TL: That's interesting. I was wondering what brought you in.

FV: But it was astronomy, I think, that brought me in. It was also history of astronomy, I was always interested in, well, where do we get these names from, where was that…and so on.

FV: Usually astronomy is thought of as an exact science or a hard science and archaeology would probably be a soft science, but the borderline isn't very definite, so really there is overlap. I suppose that we are now at the point where we say: the numbers or the figures or the measurements are telling you this…what does it mean? And what does it mean to the people in those days? When we did our work, we always said: Look, suppose we were people living in those days. What would we have there? Suppose we wanted to find the direction of the equinox, how do you find it? And that was a problem really.

TL: So, how do you think they did that?

FV: First of all, I think that the equinox is a Greek idea sort of…it comes from very late in civilizations [sic], I think. I'm not sure whether the Babylonians had a clear idea about what equinox is. I don't know; I'd have to look it up. But when we did our work, George Agius and myself, we worked in the same place, we got interested in the same topic and then we said: okay, let's go around and take the orientations of the temples. Let's do it properly. I wrote the first paper with him back in 1980 and 1981. When we started that work…I started in 1979. In the Winter Solstice I went to Mnajdra and, before that, I had taken photocopies of the plans of the temples from John Evans book.

TL: Yes, Evans 1971.

FV: That book. I took photocopies of that and I found the orientations from the plans. But then I said: wait a minute, we need to check them on site. There we found the direction of North was incorrect and when I took Mnajdra's eastern direction it was not 104° as it was on the plan (if I remember correctly). When I went on site it was about 10° out.

TL: and it should be about 90° right?

FV: Yeah. And I said: I mean, this is wrong. I went to another temple, at Mġarr, Ta' Ħaġrat, and it was correct. So, hoy, one is right and the other one is wrong? And therefore we said: look, let's take a theodolite, go there and measure it. We actually did the measurements and we got all the measurements. But then we find that this [Mnajdra South temple] is facing the equinox. Well, you take the orientation of the temple, when you take the slope of the horizon it's the equinox, I mean.

TL: Yes.

FV: But then, this is exceptional in the sense that the other temples are facing between SE-SW, alright? But then you say: ah, this is something. I mean, how can they find the equinox? What do they do? Supposing you were living there, at that time, how can you do it? And...at that time what we said was look, if they were interested in the equinox they must have found the positions of the Solstices before that. If you find the position of the Solstices then perhaps you can divide the angle, okay? And you say in-between...

TL: You're talking geometry.

FV: A sort of geometry, yeah. Although we know it's difficult to bisect the angle precisely...And also to do it on the ground, I mean to do it on a piece of paper perhaps is not that difficult. But to do it on the ground is even more difficult. And also there is the slope of the horizon.

TL: Four degree [*sic*].

FV: Exactly, there is 4° altitude. So, how can this be done? And then we also said: if we are going to look for any evidence of interest in the Solstices we shouldn't look inside the temple, we should look outside. Perhaps on the skyline, alright?

TL: Yes?

FV: So, what shall we do? A year later, it was 1981, because I remember it was a feast-day [1st May]. And when there, [as] a group probably we were about nine people. Some of us stayed on the temple and the others stood on the horizon. From the temple we could tell them: you are standing exactly on the skyline. At that time there were no mobile phones so we communicate with agreed hand-signals. Look, move back…We ask these people to move on the skyline and every number of metres, twenty or thirty metres, they had a lime solution and [used it on the rock] so we could mark it [the skyline]. We marked that and then we said let's move along that horizon, let's look about ten metres on either side and see whether we can find any [manmade] feature. And what we found was a hole. I don't know whether you know about it?

TL: I know it, I know it. Yes, I've been there, I've seen it.

FV: You saw it.

TL: Yeah, because Reuben Grima , he helped me to find it.

FV: Yes, exactly. And we found that hole. It happens to be precisely where we know they expect the Sun to rise at the Winter Solstice. The Summer [solstice] end was more difficult, because there are fields there, there are bird-trappers.

TL: Yes, I know.

FV: But later we found another hole and it was about three degrees off the Summer Solstice sunrise. Okay. And we said: ah, we found this hole and it fits, but the other one doesn't fit so we aren't sure, in any case, that the temple builders actually found the Equinox in that way. There must be some other way, which we don't know. And we left it at that, that's how we published our work.

TL: Yeah, well, that's very innovative work anyway. It's breaking...news-breaking story.

FV: What was interesting is that we predicted that there should be something and there was. So that was exciting. And, it was later then...We sent that article to *Archaeoastronomy* and to the *Journal for the History of Astronomy*. Both wanted to publish, but the *JHA* editor Michael Hoskin, told us: to publish it you need to pay so much and the figures will cost you that much. At that time I didn't have any money so we said: the American journal is not asking us for any money so we published there rather than...But then, about ten years later, Hoskin came...He came along with Georgia Foderà Serio from the University of Palermo, with whom he had done some work, I think, in Sicily. And he said: well, shall we? are you still interested? Yeah, of course I said yes; we did the work and we published also a paper. Okay. What was new then...I had noticed those drilled holes in the temple 3 at Mnajdra.

TL: The small holes?

FV: The small ones. In the stone over here, and also on that stone. But they are in this, on this stone. I had noted them and I told them, look, there are these...well, they said it's a sort of tally. Let's count the number of holes. And we counted the number of holes. We were taking notes: how many in this line, how many in that line, how many...so. And also on this [stone]. Okay? Then, in the evening, we counted them up. And they were 179 on both stones, or almost 179 on both stones.

TL: On both stones?

FV: And immediately one says 179 is close to half a year.

TL: Yeah.

FV: Close to half a year. So if we are missing some holes, perhaps we are speaking about half a year. And immediately the idea comes along that says: ah, perhaps you can find the Equinox by actually counting.

TL: ...the days, yes, yes, yes.

FV: If you count, you have so many days between there and there. If you count half the number of days you come there.

TL: But the problem is at the Solstices the sun moves very slowly so...you can't...

FV: You can't fix it precisely.

TL: Yeah. This might be our way of thinking. At that time, you know, they were not so...probably not so...

FV: Yes, yes. As I said, there are problems still. How can you fix that? But then you say, wait a minute. We found the hole near the Summer Solstice, which was 3° away. So one way to count was to say: look, the sun is here. It's still going further away. Okay? Further North. And then coming back. How long does it take? Let's say twenty days, or thirty days.

So, if it was twenty days, it's ten days away from when the Sun is here. So it is also a possibility to...I mean, so that hole could really be significant.

TL: Do you know where the hole...Is it still visible today?

FV: I don't know. I mean...the way it was found was really a fluke. It was by chance. I had given a lecture to a number of young persons. And I told them that we had found this hole, the Winter Solstice [one] and the other one is not so easy to find because there is soil, and it is on private land, etc, etc, etc. But two of them went to camp in Mnajdra. And they decided to walk along the line that they thought that there would be this hole. The farmer who had this field had cleared the soil and put it somewhere else. When they crossed the field...Well, first of all he confronted them. What are you doing here? And they told him what they were doing. And he told them, look, there's a hole here. It was really by chance, really by chance. And we went later. The farmer was very, very cooperative. We put a pole there, a wooden beam, of more than 3 metres long. And on top of it we put another piece of wood. And we put a sort of flag, so we could see it from Mnajdra.

TL: Do you have any pictures of that?

FV: Luckily I have, yes. Because otherwise it would have been sort of in imagination. But no, it's there, we recorded it.

TL: I would like to see that hole. Because I know somebody. There is a night-watchman at Mnajdra, he knows the owners there so…

FV: If he knows the owners, perhaps. It so happens that he was cooperative because the nephew of the person who owns the field is a teacher who worked with my sister, so…

TL: It stays within the family.

FV: You see, it's a small world.

TL: Yeah.

FV: But, if you can see it…

TL: Yeah, well, I'll see if in the summer…

FV: I have slides of it. If you need a photograph I'll let you have them.

TL: Yes, thank you. And of course, I'll put a reference to you. That would be very interesting.

FV: But these holes then proved to be a little more interesting because when George Agius and I looked at orientations which are perhaps aligned with the Sun, and there was only Mnajdra, [we investigated possible alignments with] the Moon and there were a number of temples [aligned] with the Moon at the extremes, and also the Stars. When we looked at the stars, we said: let's take bright stars from second magnitude and brighter. I asked somebody who I knew to send me the declinations of the stars at that time. And he sent a number of these. But, in 1979, we missed the Pleiades.

TL: Ah.

FV: Why? Because the brightest star, of the Pleiades, is about magnitude 3-3.5. And we said that's too faint. It was eta Tauri, one of the stars of the Pleiades, and we didn't realise at the time that it was part of the group, so we missed that group. So when Hoskin and Foderà Serio came along, they said: look, you missed the Pleiades, and they had the same declination as the equinox at 3,000 BC. So it could have been the Pleiades that they were interested in. And we tried various ways...the Pleiades rise with the sun, and they set with the sun in a particular time of year. They disappear of the sky [sic], because they are behind the Sun at a particular time of the year. And we thought that if we take these four: the rising, the setting, at sunset, the setting at sunrise, and the time that they are out of view, perhaps they could tell us something. And we tried to work out those numbers [and] the numbers of the drilled holes. Then, one day, I don't know how, I just said: wait a minute, suppose we are speaking about a sequence of stars: they start with the Pleiades, and then the next one, and the next one...what happens? And when you try the first one: the Pleiades, and Aldebaran (alpha Tauri) it's nineteen days. I mean, that's the first interval. And the second one fits, and the third one fits and so on. There was one that didn't fit, and we fitted it later on. So, this one, the Pleiades fitted. Nineteen days later, exactly, you get Alpha Tauri. The Pleiades rise [heliaclly], let's say, at the beginning of the year. Alright? Now, one, two, three, four, five...nineteen days later. So that's the interval. What's next? Ah, there's this star that then comes up, Alpha Tauri, which is Aldebaran, the brightest star of the constellation Taurus. Exactly nineteen days later. Okay? What's the next series? It didn't fit. I thought of another star there, not this, but...and then, I couldn't fit a star precisely. I fitted a faint star, at first. But then I say: okay let's take that star. And then this fits, this fits, this fits. The sequence fits with this calculation.

TL: Okay. So the only one is the Hyades that...

FV: But then the Hyades we said: oh, wait a minute...it's mentioned by Homer. So it was important, even for the Greeks. It could have been important for the people in those days too. And it fits. If you consider the magnitude of the group...we considered it to be 3.5, as a group. If you take 3.5 as a group then it fits there.

TL: Wow...

FV: So, that is a sequence. A sequence of what? Well, we don't know. Could be a sequence of festivals, of feasts. In our calendar we have a sequence of feasts.

TL: It's...whatever, it is a very strange coincidence anyway.

FV: A very strange coincidence. We tried to find ways to show that...The chances of really finding that sequence, fitting with the stars, I mean...And that's the project I told you that we're trying to work on. Because suppose somebody sets a random set of numbers because people said: look this was somebody who was trying to...drill, using a drill and [learning] how to use the drill. But, the lines are fairly parallel.

TL: What is this last line?

FV: That is 53.

TL: That's the 53...for?

FV: It would be...gamma Crucis.

TL: Okay, okay. So, that is a fascinating story that you have there.

FV: I need to write it.

TL: The article here, the paper here is interesting as well, but the way you put it now is even more interesting. That makes it...fascinating.

FV: I don't know whether you read the article in *Malta before History*.

TL: Yes.

FV: Because I put the whole sort of story and the work that had been done before.

TL: Yeah, yeah.

FV: And what followed and so on.

TL: You know, I've been looking at the posthole myself a lot.

FV: Alright.

TL: I'm really interested in that...At first, last summer, when Reuben Grima showed me the hole. He and the manager of Ħaġar Qim/Mnajdra, Clive Cortis, they were standing on the hole, I went back and took the angle view from the orthostat and I saw them. That was my first experience with it. So I said: okay here is something.

FV: There it is.

TL: Yeah, and then…I read your article, but I read it some time before so I didn't have this fresh in my mind so I said: okay I did that. And then, last year, on the 26th of June, the Moon rose with a declination of 24.05°, which was the declination of the Sun 5,000 years ago. I put one of the guards on the hole, with a towel in front of him…

FV: Alright

TL: …and I saw the Moon rising above him. And I managed to take a picture from there.

FV: Really?!

TL: Yes…I can send it to you.

FV: Alright, okay.

TL: And then was John Cox who helped me with it, actually. So, he said you do that and that, in a way. He wanted to have a moon illumination inside the temple, but that was at 6 or 7 o'clock in the evening so the light was not strong enough to send any light into the temple.

FV: Yes, yes.

TL: I couldn't take any picture for that, so that's what I missed. And then the Winter Solstice came and I went to take my shots again. I put up my piece of wood, and a towel so I saw that. And I saw the Sun rising about one full disk further to the North.

FV: I have a photograph of that too.

TL: I got another part into this which I will follow up as well next Winter Solstice. It's from the Middle Temple, because then I was watching the sunrise from the Middle Temple. And the Middle temple, through the portal, if you seat where on the first apse is on the left side there is this kind of altar stone where is a kind of carving on top of it.

FV: Yeah.

TL: And in the middle of there, looking straight through the middle of the doorway you hit the posthole.

FV: The posthole also?

TL: Yes, yes.

FV: I see, I see.

TL: Because I put up that pole I have to re-measure because...according to my observation like that it was hitting the posthole. But when I come back I got to measure it exactly,

[...]

TL: Because John Cox said: no, don't worry about that posthole because that's not…But I couldn't get away from it.

FV: You know, in 1979 I've got a photograph through that portal, and I went at the Winter Solstice but it was cloudy. But I've got a foggy photograph which I took in 1979…I didn't know about the posthole like that at the time, but I suspected that there was something with the Winter Solstice at that time. Then later I disregarded it.

FV: You know what? I mean, when was it, 19th, on Saturday we had a Full Moon, okay? And the Full Moon rose at azimuth 95°. And I realized [this] only at about 7:30 in the evening, I said we missed out one opportunity [sic]. So there is a difficulty also there; I mean, they couldn't have used their eyes to mark the position, exactly, of the equinox. The sun is too bright

TL: It's too bright, you can't. But what I did, well, I had a GPS, and I took a GPS reading at the entrance of the Mnajdra and then I walked due East with the GPS at 90° until I hit the horizon. And that one was just a little South of the restaurant, to the parking lot of the restaurant.

TEMPLE DIRECTIONS

TL: Do you think they were sun worshippers?

FV: I don't think so, really. I'm not convinced.

TL: No? Why did you think they placed it in that direction?

FV: Err…I don't know, really…but there must have been some…how do you call it, some intention. I mean, it wasn't by chance, definitely. There could be an indication that the early temples, the one before Tarxien are facing SE, where from Tarxien onwards there are facing SW. So they are facing rising and approximately the setting of the Sun. I mean one can think of the Sun not just rising at the horizon, sunrise, even the sun coming up, as it were, or the Sun going down. That would give you quite a latitude, as it were, sort of a width. But why they did this, I don't know…

TL: But do you see any religious, sacred, motivation…intentionality in this?

FV: I would like to see it.

TL: Yes?

FV: I would like to, but I…

TL: It's the scientist coming in?

FV: Yeah, yeah, of course my background was science. I say fortunately but...

TL: Yes.

FV: So I'm not convinced that...I mean that they were temples, yes, although some people also argue that they may not have been temples.

TL: Yes, Magli says it's circumstantial that it's called temples.

FV: Exactly. But I think that they were. So...They were dedicated to some sort of... belief. I think they are related to belief. But how? I don't know. You know the argument by the archaeologists that the inside of the temples was out of bounds for the people, no? That it was only the priests or the few that could go inside the temples. And the people would stay out. Astronomy and the orientation to the Sun and the Cosmos, sort of was important. For the people, what was important is that they were facing in a different direction, in the opposite direction. It is from where they came...their origins were from outside Malta, from somewhere in Sicily. So they were looking at the motherland, as it were, the direction of the motherland. So it's a different perspective, you see?

TL: Well, and John Cox is saying exactly the same. He's saying that the temples are facing towards Sicily from where they came. So I asked John: what do you mean by facing?

They are facing East, they are facing the other way and he said, which I didn't think: well, the entrance to the churches are in the West, and you walk towards the East, so that's the same type of concept there as well. Could be.

TL: Clive Ruggle's definition of archaeoastronomy at the ISAAC conference in Lima 2011 was, 'The study of human perception and actions relating to the sky'. Is that a definition you could use? There are many others.

FV: I...find it useful because one of the first questions one needs to ask is: were these people interested in the sky, to start with? Why should they have been interested in the sky? Is there any evidence that they were interested in the sky? And for Malta we have a bit of evidence. At least they managed to put some symbols there which look like stars, and that little [TaQadi] stone, and there are other bits and pieces so. For them, they were mostly, I think, interested in producing their food. Most of their time was spent, really, feeding themselves or finding ways of feeding themselves.

But then as they become more and more, perhaps, sophisticated, they have more time to spend and the sky was certainly part of their environment, a significant part of their environment. So, I would say yes. I mean, the sky... And then, I suppose that there would be people who would start sort of saying: well, what is the significance of the sky for us? And there are those perceptions, and those perceptions can lead to some kind of action. So yes, I would go along with that.

TL: It's probably easier to observe the Sun than the stars, that's probably why, you have to have specific knowledge.

FV: Yes, one thing I didn't emphasize about the Pleiades…is that if you want to…hum…make a temple face the Pleiades that's easy. Because they rise [in the same position] every night.

I mean they are there, just mark them. And put the temple facing them. But if you want to make the temple face the Equinox that's going to be much more difficult. Because you cannot fix the Equinox. But, with the Pleiades, it solves, sort of, the problem of how they find that direction because the Pleiades [cluster] is there.

And then it so happens that the Sun rises there also and then you get the phenomena that you see inside the temples. But then, again, what about the posthole? So there is a problem there.

TL: But so, do you think the posthole is five thousand years old?

FV: [long pause]

TL: Or do you have any, indications that it might…

FV: No, no, no. I [pause] I don't have [pause] It fits, it fits the idea that it was five thousand years ago. Now, it doesn't seem to be natural. But if you look around you, I mean, you saw for yourself, there isn't anything like it, okay? And, by the way, the other hole I told you about, the summer solstice, is clearly man-made. Very clearly. The one near the winter solstice is bit eroded probably because of weathering, etc, etc. but the other one wasn't eroded. The edges were quite sharp. So, yes, it fits.

TL: But has there been any archaeological investigation on the hole, or research and...

FV: No, no, no. A friend of mine, but he's a biologist who went along, and we said: how can you dig a hole in the rock...something [evidence of digging] must have been left in. We tried to look at the edges and see whether there are any ideas that for example they drilled around and then you would find, some remnants of the drill. He thought he could detect with his hands. But I couldn't. And then, we thought that we should ask somebody who is a geologist, to come along and have a look. That didn't happen, unfortunately. So somebody who's a specialist in rocks [karstic features] should come along and have a look at it, yes.

TL: Since Thom brought you into archaeoastronomy, have you tried to get any megalithic yards?

FV: No, no.

INTENTIONALITY

TL: So, do you think Professor Ventura that we'll ever find the reason, intentionality behind these temples?

FV: It was always my problem. What was the intention in the first place, that is the first question. Okay it's facing the equinox or facing the Pleiades; is it intentional or is it by chance? The first paper we wrote, the one I referred to, the *Archaeoastronomy* one... In 1980, 1981 we actually worked out the chances. What are the chances that, randomly, you sort of build a temple, you hit an important position of the Sun? And the chances are quite high.

TL: Yes, yes. But Mnajdra is the only one facing East.

FV: Yes, yes. Actually I explain it in this way...We have 26 axes. Suppose you take 26 matches and you just throw them up, like that, and let them fall. And you take the matchhead, as it were, as the direction. Will any of these matches hit an important position of the Sun? And, you have to consider that the width of, what do you call it, the door of the temple, the main door...is about five degrees, even more than that. Perhaps even seven degrees. Standing at the back of the temple you have a sort of angle there, okay seven degrees. And you have so many positions of the [rising and setting] sun: you have winter solstice, the summer solstice, you have the equinox. And there are positions of the moon. Alright?

And the possibility is one in four, in fact. So you have twenty six temples, it's not surprising that one of them hits an important position of the Sun. And that's putting it to chance. So, really if somebody says, look, it's just chance, he might be right. It's not enough to say: ah this temple faces East or whatever.

TL: So you think that's too pretentious to call it a solar observatory, a calendar?

FV: I think so. I think so. But it's interesting that…Look, we need to look for other evidence. We looked for the evidence; George Agius, myself and another group outside the temple because we said it has to be independent of the temple. Because if you are going to build that temple towards the equinox you have to start knowing the position of the equinox and you're going to build it in that direction. But Paul Micallef and his group looked inside the temple. Now, if you build a temple and, by chance, you hit the equinox, then it's not impossible that sunlight coming in through diagonally, that you hit other positions which then you say these are important, you see? So really, the evidence has to come about knowledge of the equinox position, or the position of sunrise and sunset before the temple is built, not after the temple is built. So, it's not inside the temple that they have to look; that's how we argued. Now whether it's a good argument or not, I don't know. You see the point. So really it is interesting.

TL: It is interesting that your theory and…the other theories looking from inside that they kind of fit.

FV: They fit, yes, they fit. And this is the problem. Is it the stars that they were looking at, or did they start with the stars and they turned to the Sun.

TL: Thank you very much Professor Ventura.

Appendix II:

A talk with Reuben Grima on: Landscape, cosmology and iconography related to Maltese prehistory.

DATE: 22/06/2011

TORE LOMSDALEN (TL): Today is the 22nd of June 2011 and we're sitting at Malta University with Dr. Reuben Grima. We are going to talk a little about things related to the Maltese temples, basically. You were senior curator for Malta Heritage for how long? Just give a little background.

REUBEN GRIMA (RG): Heritage Malta began to exist in 2003, and I was doing this job practically from the start there, later the same year. But before that, from 1992 I was an assistant curator and then a curator for the Museums Department, which preceded Heritage Malta. From around 1997 my responsibility was the management of archaeological sites, so I've been a curator for almost twenty years until I came to university earlier this year. Most of that time my focus was the management of archaeological sites.

TL: Yes, and that's what you are into today?

RG: Yes, in the broader context of what we do with the past, how we manage and relate to the past and archaeology. Of course my main research interests remain the treatment of the landscape, even the cultural construction of the landscape, particularly in Neolithic Malta.

TL: What brought you into archaeology? Why did you study that?

RG: There wasn't a moment when I said I want to do Archaeology. It's not the same for everyone, but it's something that, in my life at least, I found myself growing up with from the field I used to play in as a child where there were these rectangular cuts in the rock which have since been destroyed and probably never have been recorded. But at that time they were just 'these things there' that I now know were almost certainly Punic tombs. There are all these leftover signs and just making sense of your world, which I think is a fundamentally human trait, requires you to be an archaeologist. This is why I think archaeology is not something really on the periphery of human consciousness; to me, it seems natural for archaeology to be a central human faculty. Making sense of our world is something that everybody needs to do and that's why it's no less important than medicine or architecture, because it's something intrinsically human.

TL: Understanding the past helps us understand the present.

RG: Yes, and the life we live in. In Malta of course, this is more obvious because you live it, the archaeology is all around.

TL: Reuben, I see in these two papers I have here, one is about iconography and the other landscape. There is one thing that was striking: in the one from 2001 you used the word temples; in the landscape one from 2008 you exclude the use of [the] word temples. You use monuments, structures.

RG: Yes, perhaps.

TL: Is there a change in view?

RG: Not a change in my perspective, no. This debate is a stalemate really. Basically, for a long time now, and I'm following a trend here, many papers begin by saying the term 'temple' is a modern short-hand label which is not to be taken as an accurate description.

TL: Yes, you say that.

RG: Yes, and other people have done this before. The choice you have is either to use a slightly more convoluted label, which is less elegant like 'monumental structures' or 'monumental megalithic buildings' or 'multi-apsidal buildings'. But it's always more than one word which is always more cumbersome. There are really three ways forward: either to use the term temple with a disclaimer, as I did there. Often if it's an edited volume it's conditioned by the editor, because you don't want one paper using temple and another paper calling it something different. It's important to have some consistency within the same volume. So, in fact I don't really care very much whether I use the word temple with a disclaimer. As long as it's clear we don't really know what these buildings were for. Or to use some other term to even not accidentally risk people presuming that we know exactly what went on in these buildings. The big problem is, especially if a lay audience reads 'temple' in a narrow sense, with a modern mind where the religious is in a separate water-tight compartment from the political, the economic, the social, as it is nowadays where you have the stock exchange, the parliament, you have the temple, the night-club. And they are all very, very distinct; even the separation between religion and politics in western democracies is a fundamental tenet of democracy. Of course, we just need to scratch the surface a bit to begin to find that that separation is quite a young one. So, you know, in pre-industrial societies there's much more confusion...

TL: But, why do you think that 'temple' has been accepted as a nomination of these structures?

RG: Well, it's a label which stuck. It started to be used in the late eighteenth century and it still sticks, partly because for a long time people were thinking that these were exclusively, primarily religious buildings. And that's the danger. Nowadays we're not so confident, we're not so sure. No archaeologist has yet come up with a well-founded, alternative interpretation. People would like to come up with a model that is different from a temple, but until we have that, we just have to keep using the disclaimer to keep our options open, to keep our imagination awake on this, and primarily or certainly not exclusively religious in the sense we understand today. That's certainly in there somewhere but it's not separate and distinct. And I think I'm repeating a scheme which archaeologists hear many times; it's not an original argument at all. This is something we keep saying, even in terms of pottery description, when we call something a bowl or a spoon we are reminded that that is your modern label because of the functional attributes and associations you are carrying with you. There's a big body of literature on that.

TL: But, if you keep to the label 'temple' as a meaningful term...The temples that have an orientation towards the solstice or the equinox throw a light into the temple. You can clearly see the illumination of part of the temple inside. Then, the temple itself, is very, very small. So you think it could be a kind of a temple for the elite only, for a few people. If you take Mnajdra the masses couldn't get in, the congregation couldn't go in there; if they were inside they would destroy the illumination outside. If they were close but outside the same would happen.

RG: Restrictive access. This is also a debate in the archaeological sources where Bonanno et al., in 1993, (*Cambridge Archaeological Journal*, if I am not mistaken) puts forward the argument that during the course of the Temple Period access becomes more restricted and this is probably at the ease of priests who are using the temples to manipulate and control the masses outside. Just to keep our imagination open: to begin with, the buildings are small by modern standards; there's no space as large as this room, which isn't really such a big room by modern standards. But by Neolithic standards, if you look at the technology available you are really pushing that technology to the limits to enclose what to them must have been a vast volume, the likes of which have never been enclosed before. Given the limitations of either timber or stone beams, to have that sort of span (think of the inner apses at Ġgantija South, or Mnajdra when it was complete), to have that sort of volume is extremely ambitious. And even when they reached the technological limit of what you could roof as a single space, my suspicion is that you have a series of these spaces, the apses. Possibly part of the key to that is, rather than trying to match them up, for instance, to the shape of the statuettes (which is an argument that has been made), when you reach the technological limit of the largest space that your technology permits, you produce two of them, or three of them, next to each other. If you look at the way the temple plan develops it is really the evolution from possibly single environments to the three-apse to the five-apse—in the case of Tarxien six apses, three on each side. It is very much this tendency. The fact that you have sometimes two buildings next to each other, just creating larger and larger spaces just turns that argument on its head. There is clearly this effort to create larger and larger spaces which can create space for more and more activity. Now, of course, you can't squeeze in everyone at every single moment but there is also the dimension of time. But you don't have to be there at every single moment; you could have different people coming in at different times. Even the ritual passage through the temple could be associated with certain rites of passage, (which wouldn't surprise me at all from what we know about rites of passage in ethnographically recorded societies) it could be tied to a coming of age, it could be tied to gender, it could be only at a few episodes of your life that you have to enter the temple. So there would be plenty of room for everyone.

TL: Are you of the opinion that the equinoxes, or the illumination of the Sun at a specific area was an important manifestation for...

RG: Okay, that's a different question.

TL: This illumination of the temple is very short, actually. You talk about some minutes and certain times of the year.

RG: Agreed, and we can speculate that maybe not everybody could have been there to see it or, there to see it every year. Possibly different people would see it on different occasions. Yes, in this respect, there would be people who had more access than others, it could be gender-based, it could be age-based. It couldn't have been a free-for-all: you are right. And that is an original observation. What you are saying is exciting, it's a connection which I don't think has been made in the existing literature. There is this debate on access but bringing in the astronomical dimension makes the constrictions of space more acute. So that's good, do publish that.

TL: That's my proposal I'd like to have your thoughts on.

RG: That's good, that's very good. Not only is the space in a sense limited, because it's a confined space, but the astronomical phenomena are even more restrictive. Because they happen in part of the space, they cannot be obstructed, they happen for a short time. That makes it more special. So that does underline that, yes, some people would have had more privileged access than others, as happens today. You have to book in advance, there's a limit of forty people who can be inside [the South Temple at] Mnajdra to watch. I think Heritage Malta admits forty people for the solstices and fifty for the equinox. So yes, do write that up and publish that argument. And at Mnajdra I am inclined to believe that that is a significant alignment; observing that happening would have been very theatrical, very significant. Perhaps not so much as a calendar, not because they needed to observe that to know what the date was, but as a way of engaging with these cosmological forces as we know from so many other prehistoric and ethnographic examples.

TL: Landscape: is there a clear indication of why the temples were built where they are?

RG: My whole PhD was trying to answer that question. We might not have all the answers but I like to think that I did identify a number of patterns. There are meaningful patterns in the location of the temples in the landscape and they are really at the heart of where you'd expect people to be living. So they are very much an integral part of the lived-in landscape.

TL: You say that there are certain things that influence the site locations like the slope, elevation, water, building material, near agriculture plains...

RG: Some of those don't influence this, some do. Elevation, slope and surface geology do not seem to influence temple location. Proximity to level areas of agricultural land, access to sources of water, south-facing slopes and access to the sea are all things that they seem to care about very much, that strongly influence temple location. Probably because those are the things which matter in choosing a place to live in an agricultural society. The temples were built in these areas, in these ecological niches, with the right affordances, the right conditions to permit an agricultural community to establish itself. Which is not a surprising result.

TL: What about the orientation or alignments?

RG: There are different scales of analysis at play here. What I was just describing is at the scale of whether you build your temple in, say, Mosta or in another local place, so you've decided to be in Mosta because that is close to your agricultural land, so it's an area which, generally speaking, is an area where an agriculture society can establish itself, sustainably. Then there is a decision on a more local scale: do you choose the north-facing slope or the south-facing slope? A patch of land more level or more sloping? Looking at the sites chosen for building location they don't seem to worry much about the slope: you have temples built on quite level ground, others built on a slope. On a slope, think of Mnajdra, and on more level ground, think of Tarxien. But there is a strong preference for south, or southwest facing slopes. Only exceptionally do you find a temple on a north-facing slope. That tells you something about their priorities. In that particular north-facing temple, the temple at Ghajn Zejtuna, It is the availability of water sources which appear to have been the overriding determining factor.

So, to come to astronomy, one thing which the people writing about astronomical alignments do not mention much is this context, because if you are building on a south-facing slope the most natural direction for a building to face is downslope, or upslope depending on your entrance. You could say that is the reason why they prefer south-facing slopes, but there are other reasons: it seems to be the areas which people are living in already. From the example of Skorba we know that temples are built at the heart of living areas. So possibly that decision has not been chosen first and foremost for astronomical reasons but that is where people live in the landscape and where you build your 'temple'. So already there is this tendency for the buildings to face a generally southern direction. And this brings us to Ventura and Agius 1980, and their plot on a circle of all the axes, which are all roughly between southwest and southeast. They identified about 32 axes, and argued that if you were to randomly throw out 32 axes within this 90° range, statistically it would be unlikely not to have one which lined up with the equinox sunrise position. So their conclusion in 1980 is that one can't really say if Mnajdra is statistically significant. Now, of course with the holes on the apparent horizon, and the fact that you have also the solstice alignments, as well as all these other observations such as Mario's work on lateral alignments, and your work, I think it is an open question. It's a project under construction; this is why your work and Mario's is exciting because we don't know yet how significant it was. But testability is important, not simply saying there may be this pattern; you will convince the mainstream when you set up a testable hypothesis. Where you can say something like Ventura and Agius, with that diagram, and come up with a result saying statistically this is... I am not saying that if one does not reach that point it means it is not true, it just means it is not statistically proven. But it is a big leap forward. In an earlier paper that Mario Vassallo had written about the proportions of the building, I had suggested he apply a statistical test to his hypothesis, and he got a result that there was a less than one in twenty chance (which is accepted by archaeologists as a good rule of thumb of what is statistically significant) that the patterns he had identified could have happened accidentally. He published that, and I think no archaeologist has ever refuted it.

More recently, when we were filming David Trump for the visitor centre he told me it used to trouble him that why just the Mnajdra has this complex alignment. Could it be a fluke? But, more recently he remarked to me that none of these temples are the same, they all have their specialized functions, so it shouldn't trouble us at all is one temple is doing something that the other temples are not doing. They are all different and all special in some way, and the fact that just Mnajdra is doing that in no way reduces its significance. It may be happening at Mnajdra in a particular way that is not clear, but still staring us in the face as significant. So today he might go as far as to say that even though it's just happening here that is no reason for us not to accept that this is a characteristic of this building that possibly may be intentional.

TL: There's this paper of yours on the cosmology. What do you put into the temples and cosmology?

RG: As in many other examples that we know archaeologically and ethnographically, a built space, especially a monumental space, is meaningfully constituted. And there is this common phenomenon that meaningfully constituted spaces, even the domestic environment, are often a microcosm of the universe as we understand it. The best source-book I know for how widely attested this is is Parker-Pearson and Richards' (1994) *Architecture and Order*. But there's a lot more, since Bourdieu's study of the Berber house. Looking at these buildings, how do you go about seeing if this is the case in these temples? In the case of our temples, you can look at the devices that structure space, that characterize space. This is a technique you even find in either paintings or furniture. Applying this to the temples, first you see that different spaces are treated differently. You have the step up, for instance, between the central space and the apses. You have generally, the central space which is paved, while apses only have the beaten limestone dust. And also often the screen between one and the other are very elaborate. So that boundary has been given significance. And often you get most decoration associated with these boundaries: as sculpture concentrated around doorways, so transitions between spaces. We're getting another cue there: the transitions across these spaces are important.

TL: Yes, that's what you put under this section on crossing in this paper.

RG: In an attempt to interpret it which is not statistically provable because the sample of iconography we have is a bit too small to do statistical testing. Then we have the actual images which are of two types: portable, often three-dimensional sculpture, and the fixed, that is, the relief sculpture. The fact that it is fixed is very relevant to this argument because you know you are seeing it in the position it was designed for. Unlike a portable statue, for which the position when in use is often rather less certain.

So, looking at the low relief sculpture, and the subjects you find represented, firstly it's useful to question what we are usually told about these images. We are told that these images are of two types: the figurative, the natural representation of a picture of something from the natural world, and the abstract, which is a design, not a picture of something of the natural world. I am not convinced of that separation, because I have to approach this skeptically and critically and I have to ask why the separation? All you have to do is look at Aboriginal art and see designs which, using the same measure, would be abstract, but because we can speak to the people who made them, we can tell that one is a representation of the sea, another of a fish, or a mountain range. Once you know, you can read it. So, it is very bold to say 'this is abstract' and 'that is figurative representation'. I think we should be more prudent here. Another interesting point is the fact that the relief sculptures inhabit the same environments, treat the same boundary and, also that different subjects are never mixed in the same relief panel or even on the same block. That both things we call representation and things we call abstract follow the same rules, again, suggests they are part of the same system, which increases my suspicion that maybe they are not of two types, maybe they are of the same family. We can only speculate what the ones that have been called abstract may represent, if they are a representation. The panel in Tarxien, illustrated here, which has a thicker stem and then branches into smaller stems towards the top, is described by many as a tree-like motif.

Likewise, how would you represent the sea?

In this context, when we have a series of running spirals we simply cannot exclude that possibility. But then, the next step is looking at association: you have one block, and remember themes are never mixed, blocks are standing in for something, they are representing something. You have the block with running spirals on two of its faces, the two decorated faces because the others were hidden against the building, found in situ next to the block with fish on it. And also associated with the court. Now, most of these running spirals are associated with a specific space, a court, which is paved, which may or may not have been roofed over. What if the court was not roofed over? What if you had a wet/dry dichotomy? That is speculation; it is an argument from silence. But, hand in hand with the little iconographic evidence we have, we cannot prove it, but we cannot dismiss it.

And then we have motifs which are more terrestrial, such as the quadrupeds at Tarxien, associated with that abstract pattern which is tree-like. And that led me to think that, in the space of these temples you could possibly have a representation of the islander's world: the world as they knew it, the fundamental elements of their reality, the island and the sea that surrounded it. And that movement through the temple space is intended to invoke, record, perform ceremonies related to the way people inhabit and travel across that islandscape and the sea around it, and to give it meaning. Why? We do not know, but looking at the ritual context in so many other societies, we observe that this is a way of dealing with the things you fear, the things you do not know. And you could easily imagine that this would be a place where you would go before or after a journey, or to mediate concerns about pollution when in contact with other communities. That is an opportunity but a risk. Without imposing that idea on the Neolithic, it is undeniably a boundary; in an island the land-sea boundary is a very significant boundary in one's cosmology. And you have so many examples, both ethnographic and archaeological, where if you are in a coastal context or an island context, the boundary between land and sea becomes a key part of your cosmology and your understanding of the universe.

TL: Can you give me your definition of cosmology? What do you put in it?

RG: In the sense that I'm using it, what I mean by cosmology is the way people understand their universe. So it is not cosmology in the sense that physicists, astrophysicists use it nowadays, where it is about an understanding of outer space and the properties of that space. It is about our reality in its entirety, the world that we live and breathe. It is that definition that is commonly used in ethnographic context when trying to reconstruct the understanding of pre-industrial societies of their environment. So it is not limited to astronomy.

TL: No. There is one archaeoastronomer, Clive Ruggles; in one of the conferences he said that archaeoastronomy is the human perception and action related to the sky.

RG: Archaeoastronomy? Related to the sky.

TL: Yes, that is what he said.

RG: But did he use the word cosmology in that?

TL: No, he used archaeoastronomy.

RG: I accept that. Cosmology is a bit bigger than that because it includes that, but also it is the entire belief system: there may be an underworld in that, there may be a whole body of beliefs and practices...

TL: But all the human figurines, do you see them in relation to cosmology? Is this the fertile moon, fertile women?

RG: Some of those statues are clearly female, some of them are less clearly so. They may be female, male or androgynous or sex-less. The overtly female are more the exception than the rule. Now, in terms of tying that to this discourse, as I argue in a paper I wrote in the 2003 *Inhabiting Symbols* volume, it is interesting that the low-relief sculpture never includes the human figure. Practically all the human representations—and this is a culture which loves representing the human form, this is one of the defining characteristics—are sculptures in the round, or in high-relief. With the low-relief panels, there is flattening of the picture. High-relief is when you have a sculpture almost in the round, flattened a little, but it's still effective in 3D; you can look at it from the side as well. Think of classical Greek or a lot of Roman sculpture where sometimes even a leg could stick right out of the sculpture but the panel is still a high-relief sculpture.

I think it is important that although we have those low-relief panels and although we have all the sculpture in the round, the sculpture in the round is usually the human form (very common) and yet there is not a single example of low-relief sculpture—even in all these panels—never do you see the human form associated with them. And that lets us take the argument a little further because, possibly, what it could mean—and I speculate, I cannot prove it—but there is this pattern which must be telling us something, that representations of the human form inhabit a different plane of representation. And could that be telling us something about the function of low-reliefs? I think the low-reliefs are a bit like props in a theatre. The word props stands for properties because the props define the properties of the space. If you put up some trees in the background of a stage set, it defines the setting as a garden scene. Could the low-reliefs be the props, while statues inhabit that space almost the same way as when a living person enters that space? So are they on a different plane of reality? Are statues in the round a bit closer to actual people entering there? And often they are portable. Could these statues also be standing in for individuals? You have this in many other cults, in which a person may leave an orant figure to stand and pay homage or pray in their place. Or could they be ancestors, and their positioning in the temple related to cosmological travel?

TL: I am a little astonished that, okay, if one takes these solar alignments at solstice and equinoxes for granted, the Sun could have been very important to their universe. Still, there are very few artifacts retrieved that indicate that: you have one at Ħaġar Qim, the 'sun disk' which Fabio [Silva] (FS) doesn't think is a sun disk, you have this crescent moon and stars. So there are few artifacts...

RG: Yes, few of that sort of thing. The temples themselves are the main evidence. You will have noticed that in nothing that I have written have I built up on the astronomical alignments because I am afraid to do that myself, but I follow with great interest what Mario is writing, what you are writing.

TL: But why are you afraid of doing that yourself?

RG: Because, well, one: there are people doing a very good job at it already, and they are steps ahead of me, I cannot add to what they are saying. On the other hand, I do not have grounds to refute what they are saying, and the last thing I want to do is to try to block an idea which may or may not be right. But I always tell Mario when he is treading this fine line that I cannot tell you that I believe it, but I cannot disprove it either and you may be right, so do publish it. And usually we try to think of a testable hypothesis, and in the case of the argument on proportions I did manage to think of a way to test it and prove it to be statistically significant. But with my limitations I cannot always think of a testable way. Especially with this last paper about the winter light and the summer light. It's mind-bending, and I kept putting off giving him any feedback on it because I was very reluctant to comment on an idea I had not quite understood…

TL: But this is based on observation as well.

RG: Yes.

TL: The only thing that can destroy an observation like this, as well as Mnajdra, is if the temples were not the same five thousand years ago as they are today. This is the modern compound we are looking at here. Intentionally made for these things. Otherwise it is based on observation so it cannot be disregarded.

RG: Yes. Again, it is about different levels of confidence. I am happy to say this is possibly on the right track and extremely interesting but if you press me and say do you think that they really planned this about the summer and winter light not meeting, there I would be reluctant. I really do not know because with any building with an axis and given that the summer light and the winter light are precisely opposite each other, right?

TL: Yes.

RG: So, if you have an alignment which is approximately lined up with it, unless it is perfectly lined up, [drawing sketches and explaining] automatically this will light up this half, and this will light up the other half. It is almost a natural geometrical consequence. If it is tilted the other way the opposite would happen.

FS: But notice that, depending on the tilt, they are going to almost touch or not touch at all. I am not saying you are wrong, but...

RG: Sometimes we make observations and then build an

interpretation of them, which is what I did with the sculptures. It is interesting, it is possible, but the big leap forward is the testable hypothesis. When you say, I have seen this pattern, I predict that when we do this test, but the fact that each temple is so unique does not help, but it does not invalidate this at all. So, I am extremely interested in such, but I have to keep my feet on the ground and keep making the distinction between these different levels of provability. In my work I make a distinction between what I say about the position of temples in the landscape, which is statistically verifiable, and what I say about the deployment of images in the temple, which is not statistically verifiable and therefore more tentative. But I still make the suggestion and I put it in print and I put my name to it as an alternative reading of the evidence, which others may agree or disagree with.

TL: So, do you think we will ever find the intentionality behind what we call temples?

RG: The lesson I learned from Finn Family Moomintroll was that we do not really want all the questions to be answered. And this is one of the most difficult. So, to fully answer intentionality in its completeness…as long as there is part of it eluding us, I think it will be more interesting. We never will know every detail, and I am happy for that. I cannot even explain fully the intentionality of why we do our research. Would you want anybody to be able to explain that?

TL: No, but it would be nice to come up with a key to this but, on the other hand I agree with you, when we have that it loses interest.

RG: I do hope we have leaps forward in understanding but not the full, final answer. But of course there are so many unanswered questions that I am not worried that we will have the problem of that troll.

TL: Thank you very much Reuben; that was very interesting.

Appendix III:

A talk with Reuben Grima on: The chronology of the Mnajdra South Temple and the birth of temple culture in Malta.

DATE: 24/02/2012, Malta

TORE (TL): So, Reuben, according to you, when did they start to build the Mnajdra temples? Which are the oldest?

REUBEN (RG): It is Mnajdra generally. Everything we know about the chronology of Mnajdra is either based on typology or stratigraphy. Arguments based on typology are less reliable than arguments based on stratigraphic facts. Now, in terms of the succession between different buildings — the one fundamental incontrovertible stratigraphic fact — is that the central temple butts against, is built against the south temple, so in terms of the sequence of these buildings, in stratigraphy terms...

TL: How do you explain stratigraphy and typology in this case?

RG: Stratigraphic means based on observations of how materially one thing succeeded the other; so by looking at the succession of layers within the trenches within the building, or looking at physically how parts of the building rest one above the other, that is stratigraphic sequence, that is quite incontrovertible. By typological, I mean the form or the style.

TL: Decorations…?

RG: Yes, where on the basis of stylistic arguments, because in some other context something has been found from a certain date a particular shape, by association, by extrapolation, it may be suggested that one thing be older than the other. And this, of course, is much less reliable, and a great deal of what we say about the sequence between these temples is often more based on style and form than on stratigraphy, so we are on shaky ground. Generally speaking, for the temples generally, from the work done mainly in the 1950s by John Evans, from the calibration, the carbon dates, we learned that contrary to what was thought until then, temple building generally (and I am speaking in global terms now, all the temples on Malta) starts later than we imagined; they start somewhere around roughly 3,600 BC, the beginning of the Ġgantija phase, because the earliest pottery associated with the building of these standing structures is from that phase. So you get Mġarr and earlier phases underlying structures, but never do you find indications that there was no Ġgantija pottery around when the first walls are built; that is the broad picture. Now within that, zooming in again on Mnajdra, what is very clear—you see it on the plan, you see it when you go there—is that the central temple is resting against the lower-lying South

Temple. Beyond that, most arguments are based on typology, on the form, on the style. So even the idea that three-apsed temples are always earlier than five-apsed temples is based ultimately on typological arguments. In the case of Ta' Ħaġrat and Skorba, we have good dates for the structures; here, less so. So the main argument is that because it is three-apse, then one problem you could raise is that while the examples like the inner parts of Ġgantija or Skorba or Ta' Ħaġrat have three apses which are the same size, here we see a central apse which is much smaller, which in the classic typological sequence, is already a late revolution, a second step; so really, it is a bit odd, it is a bit different to the other three-apsed temples. So if you were to ask me if I was sure of the date of the structure I would say I cannot be fully confident on it, because the thickness of the wall is largely reconstructed in modern times, so there is very little of that left.

TL: But it seems like if you refer to Evans, if you refer to Mayr and Ashby, Mayr did not even notice the small trefoil temple in Mnajdra; it was actually Ashby who noticed it the first time, if I recall correctly; Mayr did not see, so it must have been covered or something...still it concludes that this is the oldest?

RG: And speaking of things we do not notice, there is also an even larger area of this building, larger than this temple here which—even you see in this plan—is completely left out. [See Fig. 2.4]. This is a bit like Tarxien, where you have the South, Central and East Temple, and then you have remains further east which we understand less and less. The question is whether this is the earlier part of the site? Was it still in use when this was being built, for instance, at the last stages? So that's by way of background, and in terms of the two larger structures, the Mnajdra south structures…

TL: There is an artificial foundation as well, to level it up against the south temple, right?

RG: Yes, yes…that was made and the surface you see today is also artificial in the sense that it was a modern floor put in in 1996, so you see little bits of…

TL: And the terrace there in front of the middle temple is…

RG: The terrace is circa 1949…

TL: What is this earth between the inner and the outer wall, is that filled up with earth?

RG: Yes, and most of that is undisturbed…

TL: And this is a hole in the northern apse of room 8. It is original but it doesn't go anywhere?

RG: If you look inside that hole—we're talking about the hole in the central temple in the inner apse, the hole which today looks quite rounded—if you look at old photographs, there are…I am not sure if it is one of the unpublished photographs…even the ones that are on display in the visitor centre itself, you see that hole was rectangular; today it is more rounded, it was a rectangular hole, if you look inside, you will see that on each side there are slabs, so it was almost like a cupboard space. So when you say it does not go anywhere, you mean today you do not have to look outside the building...

TL: I mean if it could be used as an oracle hole, that is where I want to...

RG: As a hole connecting with another space? It may not be connected to another space, but its purpose may be related to this void which was created by these slabs...

TL: Because this is original, this part behind it?

RG: Largely yes. There could have been some surface disturbance...but it was never torn down...

TL: So there could never have been a room behind here, another chamber behind the area for example...

RG: It does not seem like it because there is no access...we do not have any indication of it. So the sequence between Mnajdra Central or Middle temple and the Mnajdra South (and focusing more on Mnajdra South which is your main interest) you will have seen that John Evans suggests that the inner part appears older than the outer part; as far as I can recall, that argument is based mainly on the fact that the building technique is quite different, the size of the apses are quite different. In the inner part, the back part of the south temple, you have wall made almost entirely of Coralline Limestone while on the outer part the walls are almost entirely of Globigerina Limestone. There's this change in material, so consequently there is a change in building technology, and there is also a change in the proportions of the room. So looking at the plan you can see that the area of each apse, the inner apses, is less than a half of the outer apses; it was this sort of observation which led Evans to argue that there were two principal building phases, remembering that he has cut a sondage here...but that... I do not believe they were corresponding...because here you have the rock near the surface...I do not believe there is a corresponding sondage which is giving a definite date. I may be wrong, he may have cut a sondage that confirms the thickness of the walls but do check that, because if there is a sondage that gave a date, that would be one more reason to argue for that. On typological grounds it does look like a reasonable argument, and I certainly would not reject it; we know this in a similar instance where, again, the typology is suggesting that where in the South Temple the inner three apses which look very much alike, are behaving like a trefoil, which in that case are three roughly equal apses... unlike the trefoil (east) in Mnajdra and in that case in fact the outer apses are smaller than the inner apses, and the Central (middle) Temple which is believed to be a later addition, is smaller than those inner three apses, so this is why it is risky to generalise with these buildings because each one has its own uniqueness. On the basis of these observations it does look like a plausible argument. One would only be fully confident of it when confirmed with stratigraphic evidence; for instance, we get more data from the thickness of the wall which would give more firm dates, and which would continue to corroborate if that was a later structure or not. But until that is done the received wisdom (which is perfectly legitimate to follow in your argument) is that there may have been a time when there was a smaller building which would have corresponded roughly to the inner parts. Of course it would have had some sort of façade, which today we no longer can see...

TL: Yes, but then we go on to what I tried to argue: in the building, room no. 3 is the oldest one of the lot; it is older than the first three, and then building two and four are in sequence, which when based on archeology will not say it, I just based it on...

RG: Why would you say...?

TL: Well, because...the entrance itself is a temple entrance...

RG: Because it is so elaborate...?

TL: Because it is elaborate, because...it is very special...inside this room you have very special altars, the front one is a double altar, and very special...

RG: Those 'altars' also have an important structural function; they are also acting as buttresses. You can see this throughout...

TL: Buttresses…?

RG: Buttresses are…[drawing a sketch] those are called flying buttresses, those arches which are helping to support the construction. In this case you have the thrust of the apsidal walls. The horizontal megaliths over the doorway are balancing these forces and preventing the doorway from collapsing. It is double buttress…

TL: Yes, it is that, but it is very special and you have one here (which I question if it is authentic [referring to the alter between room 3 and 2]) has been there because from some pictures, this block was not there and this common supporting it was laying somewhere, so I am not sure about that. So you do not know about this and okay, this is just a theory. If they built room 3 [Fig. 5.17] first, this was closed off and they had a kind of interest in the sun, the rising of the sun, the summer solstice throws its illumination in here and the equinox does it as well.

RG: But that I find very plausible because…

TL: …because if they had an astronomical intentionality here, they did not build this finished at all; look here, we just happen to have the sun pass—I think they knew what they were doing, but then they started from scratch here in a way, I think—they might have started here; I don't say that they did not, but this is the most important room in the whole south temple, or in whole Mnajdra actually…

RG: You're making me curious, Tore, now that you say this if what we see today working here, so that the alignments with the present main entrance of the Mnajdra south, where you have those cross-jamb alignments for the respective solstices...go through the details of that for to make sense, we are hearing this without the plans of course...whether that phenomenon can...this is something you must investigate...how the inner doorway on the central axis of the Mnajdra south therefore, in this hypothetical scenario—where space one (room 1) is not built yet and the doorway which is today between space one and space two (room 2) and the entrance, the façade of the temple, it begs the question, how would these alignments...?

TL: Yes, I am going there on Sunday and I am going to check that out (room 2). I have a set of things to do here that I want to check out because I want to see what is this alignment—because this entrance here to room two, except for a lintel, one doesn't know, but the entrance itself is original, isn't it?

RG: Yes that is right, it is generally accepted. If you already have an idea and they are extending the structure and developing that idea, so each time they are building up these skills in manipulating the light and astronomical phenomena; and if what you said about this area is true, you'd have three stages to it. You won't be proved right because we do not really know if room three is earlier, but you have this hunch that this is the first experiment with light at Mnajdra; and then when room two and four are built if you can demonstrate that the same in a simpler way, the solstice cross-jamb line also works here. You must test this.

TL: I will test it; I will measure it with a compass.

RG: Even on a plan by drawing lines you can organise it—so I don't know if I have answered your question? [**RG** and **TL** drawing equinoctial and solstitial alignments on a plan (Evans 1971) from the back alter in room 2].

TL: I think about the people coming to Malta? You have written a paper on the maritime...do you have a copy of that? Could you send me...? I am interested in the Levant, the Neolithic, from the Levant, how did it come here?

RG: All across Europe, from the dates that are available, it appears that the idea of agriculture is spreading from East to West; so starting from around 10,000 years ago in western Asia, in Anatolia, from 10,000 BC, you find successively more recent dates, plenty of dates between 7,000 and 6,000; and when you come to Italy you are in the range of 6,000 BC or more recent, and Malta fits quite comfortably in that picture. It also coincides with the first colonisation of Malta, so it appears that the first colonists came here already with the knowledge of an agricultural society.

TL: So they were not hunter-gatherers, foragers?

RG: It seems far more likely that they would have been farmers, because the evidence for cultivated cereals stretches back to the first traces of human habitation in Malta. There is a logic to it, and it does make sense that on such a small island agriculture would certainly greatly improve your prospects if you want a community of any significant size to have a sound, reliable, sustainable subsistence base; you cannot feed hundreds, thousands of people indefinitely with fishing, foraging and gathering, and with agriculture you can do that.

TL: How many people lived here at the temple period?

RG: Search me! ☺ but I think Renfew's estimate of around 10,000 persons is rather high and would imagine its closer to about half that, because that calculation was based on total exploitation of the agricultural landscape. It is a more realistic figure for the medieval period when we have historic indications for a figure of around 10,000; but in prehistory possibly a more plausible model is that initially they are not exploiting the steeper slopes but are focusing on the more level areas, where you do not have the problem of erosion caused by tilling on a steep slope which you can only control by building terraces. Throughout the Mediterranean, the earliest dates we have for terracing so far are from the Bronze Age, so it is quite plausible to imagine that they would have started on the more level areas, and this argument is one I have made, I am sure you have read—things like the book...*Malta before History*. So by focusing on these more restricted areas and using that as a basis of your recalculation of population size, you'd be looking more at something like up to 6,000.

TL: About how monuments came into existence over time, do you have any opinion or theory where the temple culture came from? Was that something that grew out inside of their own population or was it external impulses to build these majestic monuments which you do not find anywhere else at that time?

RG: So specifically, the temple building phenomenon? There is nothing to have come from outside. As we were saying you have people settling the island for a thousand years before that, and even if you look at the pottery, some people describe it as part of the temple culture and for a long time it was thought that some of the earlier buildings are from the Mġarr Phase. It's quite clear that a new people is not arriving; there is so much continuity in the surrounding tradition, it is just this phenomenon that takes off very suddenly, sometime before the end of Zebbug Phase, so it does look like a spontaneous...indigenous development. John Robb's article from 2001, 'Island Identities: Ritual, Travel and the Creation of Difference in Neolithic Malta', looks at how throughout the western Mediterranean, at around 6,000 BC—you have a lot of shared elements throughout the north African coast, Spain, Italy, Sardinia, Sicily, where you have very similar decorative techniques; people are almost choosing to have the same shared idea of how to decorate pottery. If you look at the picture around 4,000 BC and more so after, you will find a much more varied mosaic of different cultures in the same region, so people are choosing to be different. In Malta, this very different way of doing things may be one expression of this, where you are asserting identity and difference. Now, why specifically megalithism, why here: there are some geological circumstances which favour it, you have a sedimentary limestone which almost offers up megaliths naturally. If you think of the inner part of Mnajdra south for instance, if you look at those rugged megaliths—in fact you do not see many pictures in books of the inner parts because it is less pretty than the outer apses—they look like megaliths that could have been natural parts of the landscape.

TL: Thank you Reuben.

Appendix IV

GLOSSARY

HELIACAL RISING[1]

Schaefer explains the heliacal rising phenomenon as; 'Many stars and other celestial bodies undergo periods of invisibility when the Sun is nearby. These periods of invisibility are bounded by the dates of the star's heliacal rising and setting. The star is first glimpsed during morning twilight on the date of heliacal rising. The apparition (or time of observability) of the star ends on the date of heliacal setting, when the Sun approaches too close to the object'.

PRECESSION OF THE EQUINOX AND THE OBLIQUITY OF THE ECLIPTIC[2]

The ecliptic is the sun's yearly path as seen from the earth. The celestial equator is the imagined projection of the earth's equator onto the celestial sphere. The obliquity of the ecliptic is known as the inclination in degrees of the ecliptic and the celestial equator. Today this angle is about 23.5°, while in the Maltese Temple Period it was around 24.05°.

---

1 Bradley E. Schaefer, 'Heliacal Rise Phenomena', *Archaeoastronomy* 11, no. 18 (1987): p. 19.
2 Kelly and Milone, *Exploring Ancient Skies*.

The precession of the equinoxes is connected to the 0° Aries Point (also known as the Vernal Point of the tropical zodiac) in the celestial sphere where the ecliptic crosses the celestial equator. The Vernal Point is not fixed, but moves westwards (backwards) along the ecliptic at the rate of approximately 1° every 72 years.

MAJOR LUNAR STANDSTILL (MJLS)[3]

A major lunar standstill happens every 18.6 years when the moon reaches its maximum northern or southern declination and is about five degrees further south or north of where the sun reaches in the sky at the summer or winter solstice. A major lunar standstill is followed by the minor lunar standstill 9.3 years later; Malville maintains that these events 'are rather underwhelming and difficult to detect', while the major lunar standstill has been detected by several prehistoric cultures around the world.[4]

HORIZON ALTITUDE[5]

The horizon altitude is the vertical angle between the horizon and a plane, stretching as far as the horizon, located at the same elevation above sea level as the observer. This is measured in degrees, from 0°, at the plane, to 90° at the zenith.

---

3 Malville, *Prehistoric*, p. 38.
4 Malville, *Prehistoric*, p. 38.
5 Fabio Silva, communication during the 'Archaeoastronomy' module, a seminar at the University of Wales Trinity Saint David (2013).

ACRONYMS OF TEMPLE ORIENTATIONS RELATED TO FIG. 3.2[6]

- **FNMR:** Far Northerly Moonrise
- **MSSR:** Midsummer Sunrise
- **MWSR:** Midwinter Sunrise
- **FSMR:** Far Southerly Moonrise
- **FSMS:** Far Southerly Moonset
- **MWSS:** Midwinter Sunset
- **MSSS:** Midsummer Sunset
- **FNMS:** Far Northerly Moonset

---

6 Cox and Lomsdalen, 'Prehistoric Cosmology'.

Bibliography

Abela, Franscesco. *Malta Illustrata: Della Descrittione Di Malta*. Malta: Paolo Bonacota, 1647. Facsimile Edition by Midsea Books Ltd, 1984.

Agius, George and Frank Ventura. *Investigation into the Possible Astronomical Alignments of the Copper Age Temples in Malta*. Malta: University Press, 1980.

———. 'Investigation into the Possible Astronomical Alignments of the Copper Age Temples in Malta'. *Archaeoastronomy* 4, no. 1 (1981): pp. 10–21.

Albrecht, Klaus. *Maltas Tempel: Zwischen Religion Und Astronomie*. Wilhelmshorst, Germany: Sven Näther, 2004.

———. *Malta's Temples: Alignment and Religious Motives*. Postdam: Sven Näther Verlag, 2007.

Antichità Fenice Nelle Isole Di Malta E Gozo. 1868.

Ashby, Thomas, R. N. Bradley, T. E. Peet, and N. Tagliaferro. *Excavations in 1908-11 in Various Megalithic Buildings in Malta and Gozo*. London: Macmillan & Co., 1913

Aull Davis, Charlotte. *Reflexive Ethnography: A Guide to Researching Selves and Others*. London: Routledge, 2008.

Aveni, Anthony. *People and the Sky: Our Ancestors and the Cosmos*. London: Thames & Hudson, 2008.

Barker, Graeme. 'Agriculture, Pastoralism, and Mediterranean Landscapes in Prehistory'. In *The Archaeology of Mediterranean Prehistory*, edited by Emma Blake and A. Bernard Knapp, pp. 46-76. Oxford: Oxford University Press, 2005.

———. *The Agricultural Revolution in Prehistory: Why Did Foragers Become Farmers?* Oxford: Oxford University Press, 2006.

Barrowclough David A., and Caroline Malone. *Cult in Context: Reconsidering Ritual in Archaeology*. Oxford: Oxbow Books, 2007.

Bonanno, Anthony, Tancred Gouder, Caroline Malone, and Simon Stoddart. 'Monuments in an Island Society: The Maltese Context', *World Archaeology* 22, no. 2 (1990): pp. 190-205.

Broodbank, Cyprian. 'The Origin and Early Development of Mediterranean Maritime Activity'. *Journal of Mediterranean Archaeology* 19, no. 2 (2006): pp. 199-230.

———. *The Making of Middle Sea: A History of the Mediterranean from the Beginning to the Emergence of the Classical World*. London: Thames & Hudson, 2013.

Brydone, Patrick. *Tour through Sicily and Mata: In a Series of Letters to William Beckford*. 1806. Reprinted by London: Forgotten Books, 2012.

Campion, Nicholas, ed. *Cosmologies*, Proceedings of the Seventh Annual Conference of the Sophia Centre for the Study of Cosmology in Culture, University of Wales Trinity Saint David, 6-7 June 2009. Ceredigion, Wales: Sophia Centre Press, 2010.

———. 'Introduction.' In *Cosmologies*, ed. Campion, pp. 1-3.

———. 'Locating Archaeoastronomy within Academia'. Paper presented at the 34th Annual Conference of the Theoretical Archaeology Group, University of Liverpool, 2012.

Caruana, A. A. *Report on the Phoenician and Roman Antiquities in the Group of the Islands of Malta*. Malta: Government Printing Office, 1882.

Cauvin, Jacques. *The Birth of the Gods and the Origins of Agriculture*. Cambridge: Cambridge University Press, 2000.

Ceschi, Carlo. *Architettura Dei Templi Megalitici Di Malta*. Roma: Casa Editrice Fratelli Palombi, 1939.

Childe, Vere Gordon. *Man Makes Himself*. Bradford-on-Avon: Moonraker Press, 1981.

Cilia, Daniel, ed. *Malta before History*. Malta: Miranda Publishers, 2004.

Clark, Daniel. 'Building Logistics.' In *Malta before History*, edited by Daniel Cilia, pp. 367–77.

Cox, John. 'The Orientations of Prehistoric Temples in Malta and Gozo.' *Archaeoastronomy* 16 (2001): pp. 24–37.

Cox, John and Lomsdalen, Tore. 'Prehistoric Cosmology: Observations of Moonrise and Sunrise from Ancient Temples in Malta and Gozo.' *Journal of Cosmology* 9 (2010).

Cox, John. 'Observations of Far-Southerly Moonrise from Ħaġar Qim, Ta' Ħaġrat and Ġgantija Temples from May 2005 to June 2007.' In *Cosmology Across Cultures*, ASP Conference Series 409. p. 344. San Francisco: Astronomical Society of the Pacific, 2009.

Davies, Douglas. 'Introduction: Raising the Issues'. In *Sacred Place, Themes in Religious Studies*, eds. Jean Holm and John Bowker, p. 5. London: Continuum, 1994.

Dicks, D. R. *Early Greek Astronomy to Aristotle*. Ithaca, NY: Cornell University Press, 1970.

Eliade, Mircea. *The Sacred and the Profane: The Nature of Religion*. Orlando: Harcourt, Inc., 1959.

Ellis, Richard. *The Photography Collection: Malta 1862–1930*. Malta: BDL Publishing, 2011.

England, Richard. 'A Space-Time Genealogy.' In *Malta before History*, edited by Daniel Cilia, pp. 408–23.

Evans, J. D. *Malta: Ancient People and Places*. Edited by Daniel Glyn, Ancient Peoples and Places. London: Thames and Hudson, 1959.

———. *The Prehistoric Antiquities of the Maltese Islands: A Survey*. London: The Athlone Press University of London, 1971.

———. 'Island Archaeology in the Mediterranean: Problems and Opportunities.' *World Archaeology* 9, no. 1 (1977): pp. 12-26.

Farrugia Randon, Stanley. *Comino, Filfla and St. Paul's Island*. Malta: P.E.G. Ltd, 2006.

Fergusson, J. *Rude Stone Monuments in All Countries: Their Age and Uses*. London: John Murray, 1872.

Foderà Serio, Giorgia, Michael Hoskin, and Frank Ventura. 'The Orientations of the Temples of Malta.' *Journal for the History of Astronomy* 23, no. 2 (1992): pp. 107-19.

Formosa, Gerald J. *The Megalithic Monuments of Malta*. Vancouver, Canada: Skorba, 1975.

Ghezzi, Iván, and Clive Ruggles. 'The Social and Ritual Context of Horizon Astronomical Observations at Chankillo'. Proceedings of the International Astronomical Union 7 (2011): pp. 144-53.

Graves, Tom, and Liz Poraj-Wilezynska. '"Spirit of Place" as Process: Archaeography, Dowsing and Perceptual Mapping at Belas Knap'. *Time and Mind: the Journal of Archaeology, Consciousness and Culture* 2, no. 2 (2009): p. 185.

Grima, Anna. 'Maltese Temple Roofing.' Private collection, 2014.

Grima, Reuben. 'An Iconography of Insularity: A Cosmological Interpretation of Some Images and Spaces in the Late Neolithic Temples of Malta.' *Institute of Archaeology* 12 (2001): pp. 48-65.

———. 'Image, Order and Place in Late Neolithic Malta', pp. 29-41. In *Inhabiting Symbols: Symbol and Image in the Ancient Mediterranean*, edited by J. B. Wilkins and E. Herring. London: Accordia Research Institute, 2003.

———. 'The Landscape Context of Megalithic Architecture.' In *Malta before History*, edited by Daniel Cilia, pp. 327-45.

———. 'Landscape and Ritual in Late Neolithic Malta'. In *Cult in Context*, eds. Barrowclough and Malone, pp. 36-40.

———. 'The Prehistoric Islandscape.' In *The Maritime History of Malta: The First Millennia*, edited by Charkes Cini and Jonathan Borg. Malta: Salesians of Don Bosco and Heritage Malta, 2011.

Hoskin, Michael. *Tombs, Temples and Their Orientations: A New Perspective on Mediterranean Prehistory*. Cambridge: Ocarina Books Ltd., 2001.

Høyerdahl, Thor. *Kontiki Ekspedisjonen*. Oslo: Gyldendal Norske Forlag, 1948.

Leighton, Robert. *Sicily before History: An Archaeological Survey from the Palaeolithic to the Iron Age*. London: Duckworth, 1999.

Lomsdalen, Tore. 'Possible Astronomical Intentionality in the Neolithic Mnajdra South Temple in Malta.' Paper presented at the European Society for Astronomy in Culture (SEAC), Portugal, 2011.

Kelly, David H., and Eugene F. Milone, *Exploring Ancient Skies: An Encyclopedic Survey of Archaeoastronomy*. New York: Springer, 2005.

Magli, Giulio. *Mysteries and Discoveries of Archaeoastronomy from Giza to Easter Island*. New York: Copernicus Books, 2009.

Malone, Caroline, Anthony Bonanno, Tancred Gouder, Simon Stoddart, and David Trump. 'The Death Cults of Prehstoric Malta.' *Scientific American* 269, no. 6 (December 1993): pp. 110–17.

Malone, Caroline, David Barrowclough, and Simon Stoddart 'Introduction'. In *Cult in Context*, eds. Barrowclough and Malone, pp. 1–7.

Malville, J. McKim. *A Guide to Prehistoric Astronomy in the Southwest*. Boulder, CO: Johnson Books, 2008.

———. 'Astronomy and Ceremony at Chankillo: An Andean Perspective'. In *Archaeoastronomy and Ethnoastronomy: Building Bridges between Cultures*, eds. Clive L. N. Ruggles, pp. 15–61. Cambridge: Cambridge University Press, 2011.

Mathieu, James R. ed. *Experimental Archaeology, Replicating Past Objects, Behaviors and Processes*. Oxford: Archaeopress, 2002.

Mayr, Albert. *Die vorgeschichtlichen Denkmäler von Malta*. München: Verlag der k. Akademie, 1901.

Micallef, Chris. 'Alignments Along the Main Axes at Mnajdra Temples.' *Journal of the Malta Chamber of Scientists* 5, nos. 1 and 2 (2000).

Micallef, Chris. 'The Tal-Qadi Stone: A Moon Calendar or Star Map.' *The Oracle, Journal of the Grupp Arkeologiku Malti*, no. 2 (2001): pp. 34-44.

Micallef, Paul I. *Mnajdra Prehistoric Temple: A Calendar in Stone*. Malta: Union Print, 1990.

Mifsud, Anton and Simon Mifsud. *Dossier Malta: Evidence for the Magdalenian*. Malta: Propprint, 1997.

Mithen, Steven. *After the Ice: A Global Human History 20,000-5000 BC*. London: Phoenix, 2003.

Neugebauer, Otto. 'An Arabic Version of Ptolemy's Parapegma from the "Phaseis".' *Journal of the American Oriental Society* 91, no. 4 (1971): p. 506.

Pace, Anthony. 'The Building of Megalithic Malta.' In *Malta before History*, edited by Daniel Cilia, pp. 19-41.

———. 'The Sites.' In *Malta before History*, edited by Daniel Cilia, pp. 19-42.

Pankenier David W., Ciyuan Y. Liu, and Salvo de Meis. 'The Xiangfen, Taosi Site: A Chinese Neolithic "Observatory"?', *Archaeologia Baltica, Astronomy and Cosmology in Folk Traditions and Cultural Heritage* 10. Klaipeda: University of Klaipeda, 2008.

Parker, Rowland, and Michael Rubinsetin. *Malta's Ancient Temples and Ruts*. London: The Institute for Cultural Research, 1988.

Patton, Mark. *Islands in Time: Island Sociogeography and Mediterranean Prehistory*. London: Routledge, 1996.

Piggott, Stuart. *Ancient Europe: From the Beginnings of Agriculture to Classical Antiquity*. Chicago: Aldine Publishing Company, 1965.

Renfrew, Colin. *Before Civilization: The Radiocarbon Revolution and Prehistoric Europe*. London: Pimlico, 1973.

——— and Bahn Paul, *Archaeology: Theories, Methods and Practice*. London: Thames & Hudson, 2008.

Robb, John. 'Island Identities: Ritual, Travel and the Creation of Difference in Neolithic Malta.' *European Journal of Archaeology* 4, no. 2 (2001): pp. 175–202.

Ruggles, Clive. *Astronomy in Prehistoric Britain and Ireland*. London: Yale University Press, 1999.

———. *Ancient Astronomy: An Encyclopedia of Cosmologies and Myth*. Oxford: Abc-Clio, 2005.

———. 'Heavenly Power in Worldly Hands: Ancient Sky Perceptions and Social Control'. Paper presented at the European Society for Astronomy in Culture (SEAC), Gilching, Germany, 2010.

Schaefer, Bradley E. 'Heliacal Rise Phenomena.' *Archaeoastronomy* 11, no. 18 (1987).

Silva, Fabio. *Cosmologies in Transition: Continuity, Innovation and Transformation in Neolithic Europe*. MA thesis, University of Wales, Trinity Saint David, 2012.

Sims, Lionel. 'Coves, Cosmology and Cultural Astronomy.' In *Cosmologies*, ed. Campion, pp. 4–28.

Skeates, Robin. *An Archaeology of the Senses*. Oxford: Oxford University Press, 2010.

Stoddart, Simon, Anthony Bonanno, Tancred Gouder, Caroline Malone, and David Trump. 'Cult in an Island Society: Prehistoric Malta in the Tarxien Period.' *Cambridge Archaeological Journal* 3, no. 1 (1993): pp. 3–19.

Stroud, Katya. *Ħaġar Qim & Mnajdra Prehistoric Temples*. Malta: Heritage Books, 2010.

Sultana, Sharon. *The National Museum of Archaeology Valletta: The Neolithic Period*. Malta: Heritage Books, 2006.

Thomas, Julian. *Understanding the Neolithic*. Oxon, UK: Routledge, 1999.

Thomson Foster, Maelee 'Orientation: A Design Determinant Is Utilized as a Means to Explore the Maltese Lower Temple of Mnajdra as a Possible Solar Calendar.' Paper presented at the International Conference on the Influence of Astronomical Phenomena (INSAP) conference, Malta, 7-14 January 1999.

Tilley, Christopher. *The Materiality of Stone: Explorations in Landscape Phenomenology: 1*. Oxford: Berg, 2004.

———. *A Phenomenology of Landscape*. Oxford: Berg Publishers, 1994.

Torpiano, Alex. 'The Construction of the Megalithic Temples.' In *Malta before History*, edited by Daniel Cilia, pp. 347-66.

Trump, David H. *Skorba*. Oxford: University Press, 1966.

———. *Malta: An Archaeological Guide*. London: Faber and Faber Ltd., 1972.

———. *Malta: An Archaeological Guide*. Malta: Progress Press, 1997.

———. *Malta: Prehistory and Temples*. Malta: Midsea Books, 2002.

———. 'Dating Malta's Prehistory.' In *Malta before History*, edited by Daniel Cilia, pp. 230-41.

———. 'Maltese Temple Cult: The Antecedents'. In *Cult in Context*, eds. Barrowclough and Malone, pp. 14-15.

Turnbull, David. 'Performance and Narrative, Bodies and Movement in the Construction of Places and Objects, Spaces and Knowledges: The Case of the Maltese Megaliths.' *Theory, Culture & Society* 19, no. 125 (2002): pp. 125-43.

Tusa, Sebastiano. *La Sicilia Nella Preistoria*. Palermo: Sellerio, 1999.

Ugolini, Luigi M. *Malta: Origini Della Civilta Mediterranea*. Malta: La Libreria dello Stato, 1934.

———. *Malta: Origins of Mediterranean Civilization*. 1934. A re-edition with a foreword and introduction by Andrea Pessina and Nicholas C. Vella. Translated by Louis Scerri. Malta: Midsea Books, 2012.

Vance, J. G. 'Description of an Ancient Temple near Crendi, Malta.' Ar-

chaeologia 29 (1842): pp. 227–40.

Vassallo, Bernard A. *Prehistoric Malta, Europe and North Africa*. Valletta, Malta: Allied Publications Ltd, 2007.

Vassallo, Mario. 'Sun Worship and the Magnificent Megalithic Temples of the Maltese Islands.' *The Sunday Times of Malta* (23 January 2000): pp.40–41; (30 January 2000), pp. 44–45; (6 February 2000), pp.36–37.

——— . 'The Location of the Maltese Neolithic Temple Sites.' *The Sunday Times of Malta* (26 August 2007): pp. 44–46.

——— . 'Ħaġar Qim's Layout Shows Yearly Movements of the Sun.' *The Sunday Times of Malta* (6 February 2011).

——— . 'Ħaġar Qim: A Leading Marker of Neolithic Time.' *The Sunday Times of Malta* (13 February 2011).

Ventura, Frank, Giorgia Foderà Serio, and Michael Hoskin. 'Possible Tally Stones at Mnajdra, Malta.' *JHA* 24 (1993): pp. 171–83.

——— . *L-Astronomija F'malta*. Malta: Pin, 2002.

——— . 'Temple Orientations.' In *Malta before History*, edited by Daniel Cilia, pp. 307–26.

Zammit, Themistocles *The Copper Age Temples of Ħaġar Qim and Mnajdra: With Plans and Illustrations*. Valletta, Malta: Facsimile Edition, 1927.

——— . *Malta: The Islands and Their History*. 2nd ed. Valletta, Malta: The Malta Herald, 1929.

——— . *The Neolithic Temples of Hal-Tarxien, Malta: A Short Description of the Monuments with Plan and Illustrations*. 3rd ed. Valletta, Malta: Empire Press, 1929.

——— , and Ing. Karl Mayrhofer. *The Prehistoric Temples of Malta and Gozo: A Description by Prof. Sir Themistocles Zammit*. Malta: Ing. Karl Mayrhofer, 1995.

Index

A
Africa, 14, 20, 219
agriculture, 14-15, 217-18
alignment, 1, 11, 72-74, 94, 104, 109, 112, 132-34, 136-38, 145, 158-62, 193-97, 205, 216
altar, 42, 111, 129-30, 136, 139, 145, 147-48, 150, 156, 159, 214
archaeoastronomy, 7-10, 66, 72, 94, 98, 103-4, 135-37, 145, 161, 181, 201
astrology, 72
astronomical observatory, 88

B
Borg in-Nadur, 58
Bronze Age, 6, 29, 31, 218
Bugibba, 92
burial, 26, 31

C
calendar, 8, 82, 84, 86, 87, 157, 174, 185, 193

Chaco Canyon (Colorado, USA), 98
Chankillo (Peru), 2
colonisation, 17-22
complexity, 137
construction (chronology), 62-63, 115, 152-53, 208-14
cosmogony, 5
cosmology, 6-7, 16, 65-69, 71, 197-98, 200-2
Crux (Southern Cross), 8-9, 71, 73
cyclic time, 80

E
Euro (coins, Maltese), 52
Egyptian culture, 84

F
fertility worship, 27
Filfla, 33, 39, 79, 107, 156

G

Ġgantija Phase (3,600–3,000 BCE), 4, 28, 41, 51, 58, 62, 136, 150, 153, 209
Ġgantija Temple, xvii, 44, 46, 74, 92, 107, 134, 136, 158, 191, 210
Ghan Zejtuna, 195
Ghar Dalam Phase (5,200–4,500 BCE), 19
Gozo, xvii, 3, 9, 18, 19, 21, 25, 26, 27, 28, 44, 46, 58, 73, 92, 114
Ħaġar Qim, xviii–xx, 33, 34, 38, 48, 58, 70–74, 78, 80, 92, 134, 136, 156–58, 203

H
Hawkins, Gerald, xix,
heliacal rise, xxi, 82–85, 221
History of Malta, xvii–xxi
Holocene, 17, 18
horizon astronomy, 8
horizon calendar, 2
Hypogeum (Hal Saflieni), 29, 50, 92

I
Ice Age, 14, 20
iconography, 198–200, 203
identity, 68, 220
INSAP, 78
intentionality, 1, 11, 102, 138, 142–44, 155–62, 179, 183–85, 197, 206, 215

J
Jupiter, 98–100

K
Kon-Tiki, 98
Kordin III, 92

L
landscape, 36–37, 69, 102–4, 144, 188, 194–95, 200–1, 206, 218, 220
Levant, 14, 217
limestone, 68, 220
    Coralline, 39, 46, 213
    Globigerina, 37, 41, 50, 116, 213
lunar cycle, 112

M
major lunar standstill, 85, 98, 222
maritime connectivity, 6
megalithic temples, 3
menhirs, 15
Merleau-Ponty, Maurice, 104
Mesolithic, 14–15
Mġarr Phase (3,800–3,600 BCE), 27, 58, 209, 219
Mġarr Temple, 44, 50
migration, 17
Misqa Tanks, 38
Mnajdra, origin of name, 33
Mnajdra East Temple, xxi, 32, 39–41, 62, 74, 107–8, 134, 152–53, 156

Mnajdra Middle (North) Temple, 32, 37, 41–46, 54, 63, 85, 96, 101, 107, 108–12, 134, 136, 138, 139–40, 148–49, 151–53, 157
Mnajdra South Temple, xx–xxi, 1, 10, 11, 32, 37, 46–62, 72, 85–87, 92, 93, 96, 100, 107, 112–34, 136, 138, 141–45, 148–49, 151–53, 158–60, 215
Mnajdra Temple complex, 13, 29, 62–63, 80, 134, 152–53, 161, 191, 195, 209, 212–13
monuments, 16, 23, 188, 219
Mount Etna, 18

N
Nabta Playa (Egypt), 20
Naffara Hill, 107
Neolithic diffusion, 13–16
niche, 41, 42, 45, 52, 54, 110, 129, 131–32, 145

O
objects, 25, 56
oracle holes, 128–30, 137, 142–43, 212
orientation, xx, 11, 16, 72–74, 78–80, 94, 104, 107, 110, 112, 114, 129, 131, 136, 145, 156–58, 162, 194
Overton Down (England), 97

P
Paleolithic, 14

parapegma, 82
passage graves, 15
Phoenicians, xvii, 18–19
Pleiades, xxi, 82, 84–85, 172–73, 182
pottery, 22, 25–26, 31, 37, 39, 41, 51, 56, 58–59, 69, 78, 150, 209, 219
precession (of equinox), 73, 221–22

R
radiocarbon dating, xxi, 23
randomness, 10
Red Skorba Phase (4,400–4,100 BCE), 24
ritual, 29, 80, 137, 142–43, 157, 192, 200
ritual sacrifice, 48
roof/-s/-ing, 35, 44, 50, 70, 140–42

S
sacred buildings, 4–5
shrine, 5, 25, 39
Sicily, 20, 21, 22, 26, 79
Sirius, 73
Skorba Temple, 19, 37, 69, 92, 134, 136, 195, 210
spatiality, 104, 197–98
spiral/-s, 67, 71
stone circles, 15
Stonehenge, xix, 2, 98
stratigraphy, 208–9, 214

T
Ta' Ħaġrat Temple, 27, 92, 134, 136, 142, 210
Tal-Qadi stone, 8, 71, 78, 92, 181
tally, 82, 157, 169
Taosi (China), 2, 3,
Tarxien Phase (3,000–2,500 BCE), 4, 28, 31, 45, 50, 51, 58, 62–63, 110, 136, 149, 153
Tarxien temple, 8, 93, 134, 136, 191, 195, 210
temple carvings, 67, 68
Temple period (3,600–2,500 BCE), 4, 23, 24, 29, 85, 100, 112, 118, 142, 156, 190
temples, term, 180, 188–90
Thom, Alexander, xix
tombs, 27
torba, 50, 66
trefoil, 34, 40, 41, 79, 213

U
UNESCO, xviii

X
Xemxija Heights, 26
Xagħra Circle, 93

Z
Zebbug Phase (4,100–3,800 BCE), 26, 37, 58, 219
zodiac/-al circle, 71

SOPHIA CENTRE PRESS

www.ingramcontent.com/pod-product-compliance
Lightning Source LLC
Chambersburg PA
CBHW042124100526
44587CB00026B/4170